A Garland Series

The Modern Chinese Economy

Edited by
Ramon H. Myers
HOOVER INSTITUTION

The Livestock of China

Ralph W. Phillips
Ray G. Johnson
Raymond T. Moyer

Garland Publishing, Inc., New York & London
1980

Bibliographical note:
this facsimile has been made from a copy in
the Yale University Library.

Library of Congress Cataloging in Publication Data

Phillips, Ralph Wesley, 1909–
The livestock of China.

(The Modern Chinese economy)
Reprint of the 1945 ed. published by the U.S.
Govt. Print. Off., Washington, which was issued as
no. 2249 of Department of State publication,
and as no. 9 of Far Eastern series.
Bibliography: p.
1. Livestock—China. I. Johnson, Ray George, 1902– joint
author. II. Moyer, Raymond Tyson, 1899– joint author. III. Title.
IV. Series: The Modern Chinese economy. V. Series: United
States. Dept. of State. Publication ; 2249. VI. Series: United
States. Dept. of State. Far Eastern series ; 9.
SF55.C6P5 1980 338.1'76'00951 78-74304
ISBN 0-8240-4286-7

For a complete list of the titles in this series,
see the final pages of this volume.

All volumes in this series are printed on acid-free,
250-year-life paper.

Printed in the United States of America

THE LIVESTOCK OF CHINA

By

RALPH W. PHILLIPS
SENIOR ANIMAL HUSBANDMAN IN CHARGE OF GENETICS INVESTIGATIONS
BUREAU OF ANIMAL INDUSTRY, U. S. DEPARTMENT OF AGRICULTURE

RAY G. JOHNSON
HEAD OF THE DEPARTMENT OF ANIMAL HUSBANDRY
OREGON STATE COLLEGE

RAYMOND T. MOYER
ACTING HEAD OF FAR EAST DIVISION, OFFICE OF FOREIGN AGRICULTURAL RELATIONS
U. S. DEPARTMENT OF AGRICULTURE

UNITED STATES GOVERNMENT PRINTING OFFICE
WASHINGTON : 1945

DEPARTMENT OF STATE

Publication 2249

FAR EASTERN SERIES 9

For sale by the Superintendent of Documents
Government Printing Office, Washington 25, D. C.

PRICE 30 CENTS

Foreword

In the spring of 1942 the Chinese Government accepted an offer from the United States Government to send a number of American technical specialists to assist in the Chinese war effort. The Chinese Foreign Office indicated that there was urgent need for technicians in several fields, including various branches of agriculture. One was to be an animal breeder and another a specialist in range management.

The Department appointed more than 25 technical specialists for service in China, one of whom was Dr. Ralph W. Phillips, an animal-breeding specialist of the Department of Agriculture and editor of the widely read *Journal of Animal Science*. Another was Prof. Ray G. Johnson, head of the Department of Animal Husbandry at Oregon State College, a ranch owner of long experience in range management. Both spent the year 1943 in China.

This booklet is the result of their period of work with the Chinese Government. In its preparation, assistance was provided by Dr. Raymond T. Moyer, for many years director of the Minghsien College agricultural experiment station at Taiku, Shansi Province, in North China, located during the present war at Chintang, Szechwan. Dr. Moyer conducted a number of animal-breeding experiments here during the years 1931–42. He is now in the Department of Agriculture, Washington.

The Department of State is publishing this report for distribution to agricultural experiment stations and schools, both in China and the United States, in the hope that the publication will make more widely known the findings of the two travelers and will contribute to the closer collaboration of agricultural scientists in the two countries.

DEPARTMENT OF STATE
October 1944

Contents

Introduction

THE development of animal husbandry in any country is related intimately to the mode of life of the people. The mode of living is in turn closely associated with geographic and climatic conditions. There is a wide variety of geographic and climatic conditions in China and there are two main modes of living among the rural populations, represented by the settled farmers of the thickly populated portions of the country and the nomadic and seminomadic peoples of the outlying regions of the west and northwest. In addition to variations in the environment, there are variations in the racial stocks, including the Chinese, Mongolian, Tibetan, and others.

The Chinese farmer is a tiller of the soil and has made much progress in the development of terracing, irrigation, and other methods of improving the producing capacity of the soil. But he appears not to be so skilled in animal husbandry.

In the areas of settled agriculture, where a large proportion of the total livestock is actually raised, two basic difficulties are present. First, pressure of population upon supply of foodstuffs has made it impossible to feed to livestock anything like the proportion of sound grain—which could also be used directly as human food—that is fed in countries where a livestock industry is highly developed. And second, farmers have operated without benefit of knowledge of heredity, nutrition, and health.

In spite of handicaps which these difficulties have imposed upon producers of livestock, certain achievements can be pointed to: The forms of livestock raised in China are varied, and the different forms are generally found in the region to which they are best adapted. The breeds used, although unproductive from a Western viewpoint, are able to survive and to contribute products acceptable to the Chinese taste even though raised under a poor environment.

The forms of feeding materials available to Chinese farmers in the different regions are utilized practically to the limit, and the amount of livestock products produced under the existing environmental conditions is surprising. The existence of a livestock industry in China has resulted in a greater rather than a smaller total quantity of foodstuffs being produced because of the basis of operation, which is quite unlike that in the United States and many other countries.

The nomadic and seminomadic groups of the inner-Asian grassland areas are largely dependent upon livestock for their existence and therefore have a greater natural interest in animal husbandry than the more settled farmers. But most of these peoples live in areas where the winter climate is rather severe. Relatively little land has been developed for crop production, and the areas upon which crops can be grown successfully are limited. Few attempts have been made to utilize this land for the production of supplemental feed for the livestock.

Active livestock-improvement programs have not been undertaken in either the farming or pastoral areas, and knowledge of improved methods of breeding, feeding, and management is very limited among the rural populations. Some native types of livestock are meeting the needs of the country reasonably well, but most types are relatively unimproved and their production of meat, milk, wool, or work could be increased. In many areas the management practices and available feed supplies are such that improved types could not be raised satisfactorily at the present time. Some areas are stocked with more animals than the available forage can adequately support and many grassland areas are suffering from overgrazing.

There are relatively few areas where the numbers of livestock should be materially increased until management practices are materially improved, or unless land that is now used for production of human food is taken out of cultivation for that purpose. China cannot afford to take much, if any, land out of use for human food production unless marked increases in production of crops per unit of land can be brought about.

A general interest in livestock improvement has not yet developed, and this is naturally reflected in a scarcity of trained animal-husbandry specialists to assume leadership in this field. A few men have gone to other countries for advanced training and some training is now being given in China, but much remains to be done to meet the country's needs. Owing to the limited number of trained workers and the difficulties of transportation and communication, little information has been compiled concerning Chinese livestock and the conditions under which they are produced. The object of this publication is to present a summary of the existing information so that it will be available for the use of Chinese workers and others interested in the development and utilization of Chinese livestock and their products.

Geography, Climate, and People

The variability in geography, climate, and racial origin of the people of China has already been indicated. Space will not permit detailed discussions here, and other authors have already covered these subjects in detail. Cressey (1934) presents a comprehensive survey of the land and its people, including geographic and climatic aspects, and Latourette (1934) presents a detailed account of Chinese history and culture.

China is a land of mountains. Level land is found along some river

FIGURE 1.—Map of China showing the major agricultural areas. *A*, Rice region. *B*, Wheat region. Kaoliang and millet are also important food grains in portions of this region. *C*, Grasslands and desert area.

FIGURE 2.—Kweichow is a mountainous province, with rather heavy rainfall and little tillable land. *Top:* Coarse grass grows on the mountains that are barren of trees. View near Meitan, Kweichow. *Bottom:* The Burma Road winds down the steep head of a narrow valley. There are over 30 sharp U-turns on this slope. Note the flooded rice fields along the various shelves on the hillsides.

valleys, in delta regions, and a few other areas. The largest of these level areas is the compound delta that reaches from northern Chekiang Province northward into Manchuria. There are two large inland basins in Kiangsi and Hunan Provinces, around Poyang and Tung-ting Lakes, respectively. There are a few alluvial areas south of the Yangtze Valley, including small deltas in the vicinity of Swatow and Canton. West China has limited level areas in the alluvial fan of the Chengtu plain and the plateau of Yunnan. Rather large areas of level or rolling country are also located in Tibet and Mongolia, being mostly

FIGURE 3.—Rice is the main crop in southern China and terraced fields are seen wherever they could be built. Some are flooded throughout the year while others are dry during the winter. *Top:* Flooded rice fields in Szechwan, about 40 miles west of Chungking. *Bottom:* Rice fields and manure pits near Yungchang, Szechwan. Some of the fields have been drained and are producing a crop of winter vegetables. The manure pits are of much better construction than most that are in use. All available animal and human feces are stored in pits and, as they disintegrate, liquid is dipped out and spread over the fields.

FIGURE 4.—*Top:* Terraced hillsides in mountains near the Taiyuan plain in Shansi. *Middle:* Terraced hillsides at the edge of the Taiyuan plain. *Bottom:* A view of the Taiyuan plain, near Taiku, Shansi.

FIGURE 5.—Erosion in the loess soil of the Shouyang district in Shansi. Trees in the central foreground are a pear orchard.

in the nature of high plateaus, and there are some level areas in Sinkiang. The remainder of the country is mountainous, much of it being quite rugged. The Tsingling Mountains and associated ranges on the southern border of the Yellow River Basin form the natural boundary between the humid rice area of the south and the drier area of the north. Buck (1937) classifies the area south of these mountains as the rice region and the area north of them as the wheat region, these being the most important food crops grown in the two areas. Some areas in northern China and especially in the more recently developed areas of Manchuria have considerably larger farms and fields than the averages given below for the rice and wheat regions. The grasslands lie to the north, northwest, and west of these major crop areas, and in the more mountainous portions of crop areas.

China's farming area is for the most part a land of small fields. Where the country is level they are usually square or rectangular. In the hill country of the south the rice fields are terraced and follow the contours, and the landscape is very irregular. Terracing is also used extensively in the northern portion of the country, but since flooding of the land is not practiced here, as it must be in the rice area, the terraces are not so level and are not built along natural contours so carefully. Irrigation is practiced extensively in many plains and valleys. Open, unfenced fields are the rule. Individual farms are small. Buck (1937) surveyed a large number of farms in the rice and wheat regions and found the average acreages to be 3.09 and 5.63, respectively. Even these small areas are broken up so that the average number of parcels of land per farm was 5.6, the rice and wheat areas being nearly identical in this respect. Everywhere in the farming regions there is an oversupply of people and an under-

FIGURE 6.—*Top:* High plateau country in Suiyuan (part of Inner Mongolia) with Mongolian shepherds watching their flocks. *Upper middle:* A walled village in Suiyuan, typical of the open plateau country of the northern portion of China. *Lower middle:* Scene at the edge of the plateau country in Suiyuan, looking from a lower elevation toward the edge of the plateau. *Bottom:* Farming area in a valley in the southern portion of Suiyuan.

FIGURE 7.—A view near Lanchow, Kansu, with the Yellow River in the middle background. Contour trenches in foreground have been planted to Chinese elm, locust, and tamarix in an effort to reforest the hillsides and stop erosion.

supply of tillable land, and every available plot is pressed into use for the production of human food. Obviously there is little room for livestock other than the essential workstock in the densely populated areas of China.

The grasslands, for the most part, are made up of mountainous and high plateau regions. The nomadic and seminomadic peoples in these regions herd most of their livestock on the higher mountainsides and plateaus during the summer and bring them into the lower valleys in the winter months. This is especially true of the grasslands in Mongolia and Manchuria. Except for the very high regions of the Tibetan grasslands, the climate in this area is somewhat less severe than in Mongolia and Manchuria. In all these areas, however, the natural feed supply is low during the winter and little feed is grown and stored for livestock. Consequently, animals are usually in poor condition at the end of the winter season and heavy death losses are frequently incurred. Much of Sinkiang is at a lower altitude than the Tibetan and Mongolian plateaus and the temperature is gener-

ally more favorable to successful livestock production. In all the grassland areas, overgrazing and other mismanagement of the grazing areas present problems that are in urgent need of attention. Lack of communication and transportation between these grassland areas and the farming areas are among the factors that have prevented development of a balanced economy such as is found between the western grazing areas and the eastern portion of the United States.

Racially, the population of China is of diverse origin, particularly in the border provinces. Latourette (1934) states that there is much uncertainty as to the origin of many of the groups. Many invasions of the fertile lands of North China by non-Chinese stocks from the northeast, north, and west undoubtedly left their marks on the population of that region. The Koreans have intermingled with other stocks to a limited extent, especially in Manchuria. Similarly, there are many types in the southern portion of the country, owing to mixing of the Chinese with native tribes and with other people of southeastern Asia. In addition to the Chinese, the predominant groups are the Tibetans, Mongolians, and a racial mixture among Chinese Moslems which includes at least Turkish, Mongolian, and Arab strains mixed with Chinese. In addition, there are many tribes that are numerically small, such as the Kazakhs of the extreme northwest and the Miao of the mountainous regions of Kweichow and adjacent territory.

The Chinese are the predominant racial stock in the densely populated farming areas. Less aggressive native tribes, such as the Miao, have been pushed back into the poorer mountain regions. Chinese farmers have pushed into the more fertile lands at the border of the Tibetan and Mongolian areas and many have migrated north into Manchuria, and in these border regions there is competition between the pastoral and agrarian economies. Moslems are found in largest numbers in Chinghai, Kansu, Ningsia, and Sinkiang Provinces.

FIGURE 8.—The valleys and hills of eastern Chinghai are rather densely populated, and most of the tillable land is used for crops. Upland regions are occupied by Tibetans and their herds and flocks. The higher regions of the central, southern, and western portions of the province have few people. *Top:* View from Lauye Mountain, north of Sining, overlooking the Datung River Valley. Note the extensive terracing on the mountains in the background. *Middle:* Terraced hillsides west of Sining. Higher mountains in distant background are too rugged even for grazing. *Bottom:* Women and children of a Tibetan camp, in one of the high valleys. Note the yaks tethered for milking. The heavy silver ornaments on the backs of some of the women constitute part of the family fortune.

615999—45——2

FIGURE 9.—Sikong, in the eastern portion of the Tibetan highland country, has a rugged landscape, with much good grazing land. *Upper left:* Steep mountains and sparse vegetation at an elevation estimated at about 18,000 feet. *Upper right:* A view near Taofu. *Middle left:* High grasslands north of Lihua. Note the Tibetan yak-hair tents in the valley and the yaks grazing on the hillsides in the right background. *Middle right:* Excellent grasslands near Lihua. *Lower left:* Scene along the Yalung River. The bare spot in the center is a potato field and the two light spots above are corn fields. Farmers burn brush on the hillsides, plant buckwheat in the ash, and in the following year plant corn, potatoes, or beans. If any soil is left, one of these crops is planted in the third year. *Lower right:* Pea and barley fields in a narrow valley area at the Paomei Experiment Station.

FIGURE 10.—Living conditions are rugged for both man and beast in Tibet. *Top:* Tibetan sheep near Dochen. *Middle:* Yaks grazing on lower mountain slopes. *Bottom:* A view in Tangla pass, between northern India and Gyantse, Tibet.

FIGURE 11.—Tibetan valleys are used to grow grain and other crops for human use and for quartering livestock during the colder weather. *Top:* Herder's shelter in a valley near Samada, Tibet. Cakes of dried manure for fuel are stacked in the foreground. Note the erosion on the hillside in the background. *Bottom:* A threshing floor in Tibet, with ponies and cattle being used to tread the barley.

Livestock Numbers

A reasonably accurate estimate of the livestock population of a country is difficult to obtain, even with the most elaborate census-taking organization. Under Chinese conditions, with a very low level of literacy in rural areas, poor communications, and either many small units in densely populated areas or isolated units in grassland

TABLE 1

ESTIMATED NUMBERS OF LIVESTOCK ON FARMS IN VARIOUS PROVINCES OF CHINA IN MARCH 1937, EXCLUSIVE OF SIKONG, TIBET, SINKIANG, OUTER MONGOLIA, AND MANCHURIA

(Numbers Are in Thousands)

Province	Type of Animal										
	Horses	Donkeys	Mules	Cattle	Water Buffalo	Sheep	Goats	Swine	Chickens	Ducks	Geese
Chahar	45	104	46	59	....	444	169	129	793	24	5
Suiyuan	151	112	41	301	....	1,411	539	282	1,040	2	1
Ningsia	17	44	8	39	....	739	225	52	186	6	2
Chinghai	50	79	26	89	....	359	150	74	312	3	2
Kansu	151	681	160	802	1	2,625	1,646	984	3,212	229	53
Shensi	128	359	131	914	7	375	657	967	2,747	99	7
Shansi	118	562	263	513	3	2,198	1,660	528	3,874	45	5
Hopeh	317	1,194	763	1,167	15	789	1,052	3,742	13,480	480	50
Shantung	325	2,026	744	2,580	11	1,089	969	3,506	21,355	2,298	355
Kiangsu	109	712	99	1,267	906	329	1,235	5,018	18,394	6,190	709
Anhwei	203	653	178	1,037	792	135	407	2,802	15,832	3,574	1,447
Honan	487	1,900	676	3,139	192	889	1,529	3,187	18,370	2,263	262
Hupeh	245	428	153	1,875	924	59	986	3,931	21,458	2,481	309
Szechwan	89	31	67	824	1,999	110	1,482	8,177	15,858	5,304	757
Yunnan	342	88	202	485	542	184	611	2,761	5,882	844	163
Kweichow	188	6	28	568	601	32	245	1,422	3,748	874	112
Hunan	37	10	12	1,369	1,482	9	387	5,030	16,310	6,568	396
Kiangsi	53	12	14	1,745	827	3	157	3,800	17,526	4,685	1,028
Chekiang	3	4	5	890	314	[a] 613	739	2,718	16,444	3,344	879
Fukien	6	6	2	354	276	8	286	1,757	6,002	2,321	369
Kwangtung	45	2	4	1,579	1,299	11	439	5,187	22,059	7,424	2,203
Kwangsi	151	5	2	1,485	1,383	...	174	3,650	16,968	6,338	402
Total	3,260	9,018	3,624	23,081	11,574	12,411	15,744	59,704	241,850	55,396	9,516

[a] Probably too high. Based on Chinese National Agricultural Research Bureau, *Crop Reports*, vol. VII, nos. 5 and 6, June 15, 1939.

areas, the taking of even a rough census is difficult and in some areas practically impossible. Rough estimates of the numbers of livestock have been obtained by the Chinese National Agricultural Research Bureau, and those for 1937, published in their crop reports in 1939, are summarized in table 1. Owing to the difficulties of obtaining even rough estimates, these figures have limited usefulness but they give at least some measure of the relative concentration of various types of livestock in each province. Sikong, Tibet, Sinkiang, Outer Mongolia, and the Manchurian Provinces are not included in these estimates.

No data are given on numbers of camels in the Inner Mongolian provinces and in Kansu and Chinghai, where considerable numbers of these animals are used. Estimates obtained from government officials in Ningsia indicate that there are approximately 200,000 in that province.

Data are not available on the numbers of livestock in Sikong and Tibet. Some estimates are available for Sinkiang, collected by the government of that province in 1941, and these are summarized below:

Type of Animal	*Estimated Number*
Horses	989, 195
Donkeys and mules	307, 514
Camels	90, 740
Cattle	1, 208, 438
Sheep	9, 431, 671

It seems probable that the figure for sheep includes both sheep and goats. The reports mention some swine in the area of Yilee but do not indicate that any are found in other sections.

Kislovsky (1938) gives the following data on livestock numbers in Outer Mongolia, based on a census taken in 1930:

Type of Livestock	*Estimated Number*
Horses	1, 550, 400
Camels	480, 600
Cattle	1, 380, 600
Yak	449, 600
Pien niu [a]	40, 700
Sheep	15, 574, 500
Goats	4, 075, 700

[a] Hybrid between native bull and yak cow, or the reciprocal cross.

No mention is made of donkeys, mules, swine, or poultry.

The numbers of livestock in Manchuria, as given in the Japan-Manchukuo Yearbook for 1937, are listed below. The figures are for 1935, except those for sheep and goats which are for 1936 and from the 1939 Yearbook.

Type of Livestock	*Number*
Horses	1, 808, 896
Donkeys	620, 971
Mules	567, 729
Camels	10, 361
Cattle	1, 276, 769
Sheep	1, 680, 545
Goats	751, 587
Swine	5, 911, 167
Chickens	9, 860, 552

Ross (1936) points out that there was considerable discrepancy in the estimates of sheep in Manchuria. The research department of the South Manchurian Railway Company gave the number as 2,720,000, of which 1,959,000 were in the Mongolian provinces, while the Mongolian Affairs Department gave figures of 2,000,000 and 1,087,000, respectively. Using available figures on wool exports and assuming that 25 percent of the wool produced was consumed locally, Ross arrived at an estimate of about 10,000,000 sheep in Inner Mongolia and Manchuria. Of these, he states that about 10,000 in Manchuria and 1,000 in Inner Mongolia are of Merino and crossbred type, the remainder being fat-tailed Mongolian sheep.

Granting that there may be considerable inaccuracy in the estimated numbers of livestock, it is obvious that China has a large livestock population and that it must be considered in any attempt to improve the productivity of the agricultural industry of China.

Horses

Types

The horses of China may be divided into three main types, Mongolian horses, South China ponies, and Sikong horses or ponies. Several subtypes are recognized among the Mongolian horses and there is considerable variation within all three types. More adequate information must be obtained before an entirely satisfactory classification can be made on the basis of anatomical and size differences.

The Mongolian horse is found throughout the grassland areas of China and many of them have been taken into adjacent farming areas in the northern portion of the intensely cultivated section of China. In general, Mongolian horses are rather light-boned. The head is carried quite low and in some animals is below the level of the withers

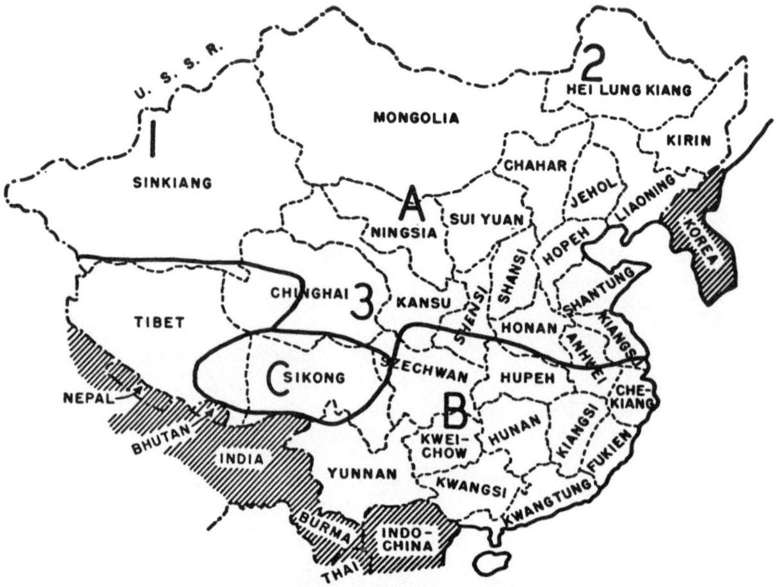

FIGURE 12.—Distribution of various types of horses in China. *A*, Area in which Mongolian horses are produced. *1*, Home of Yilee horses. *2*, Home of Hailar horses. *3*, Home of Sining horses. *B*, Area in which South China ponies are found. *C*, Area in which Sikong horses are produced.

when walking. The neck is light and frequently drops off in front of the withers. Shoulders are straight and withers are low. Usually the back is somewhat low and the croup is steep. Leg bones are small, and crooked rear legs or sickle-hocked animals are the rule. The sickle-hocked condition is not looked upon as a serious defect, or even as a defect at all, by many people in the grasslands. The hoofs are of good size and are reasonably well shaped and of good wearing quality. "Hog-backed" animals (backs slightly high in the middle) are also seen among the Mongolian horses. There is wide variation in color, most of the colors common to horses being seen frequently. A white blaze or spot on the face is not favored since it is supposed to be a bad omen. Chinese Army figures indicate that the Mongolian horse is about 130 cm. in height at the withers. Twelve mares from Suiyuan and 25 from Ningsia, that were 4 years old or over, at the Ministry of Agriculture and Forestry's First Horse Breeding Station averaged 128 and 128.4 cm., respectively, in height. Animals in these groups varied from 120 to 137 cm. (approximately 12 to 14 hands).

Kislovsky (1938) describes the horses of Outer Mongolia as little more than ponies, averaging about 126.9 cm. at the withers and being of a rather compact build. The mane and tail are abundant. Chestnuts are often lacking on the hind feet. The color is described as being quite variable, all known colors being seen, but bays and grays were observed somewhat more frequently than any others. The horses of central Mongolia are described as being smaller (height, 123 cm.) than those in western Mongolia. There may be some admixture of Kirghiz blood in the latter type.

The terms Sining horse or Nanfan horse are applied to a type of Mongolian horse found in southwestern Kansu and eastern Chinghai. These horses are a little larger than horses of surrounding areas and are much prized as saddle and pack animals. Many of them are used along the pack trails into Tibet. Some of the animals found in the Kansu-Chinghai border area are well muscled and sufficiently heavy and low-set in conformation to classify as small farm chunks. The faults commonly seen in Mongolian horses are also found in Sining horses.

A number of subtypes have apparently arisen from crossing Mongolian horses with somewhat larger horses from adjacent areas in Russia. These include the Yilee, Hailar, Sanpeitze, and possibly others. The Yilee is the most noted of these subtypes and is reported by the Ministry of War to average about 138 cm. in height. They are light in build, being of saddle rather than draft type. In conformation, they have many of the faults common to Mongolian horses but in general are of better conformation and of somewhat stronger build.

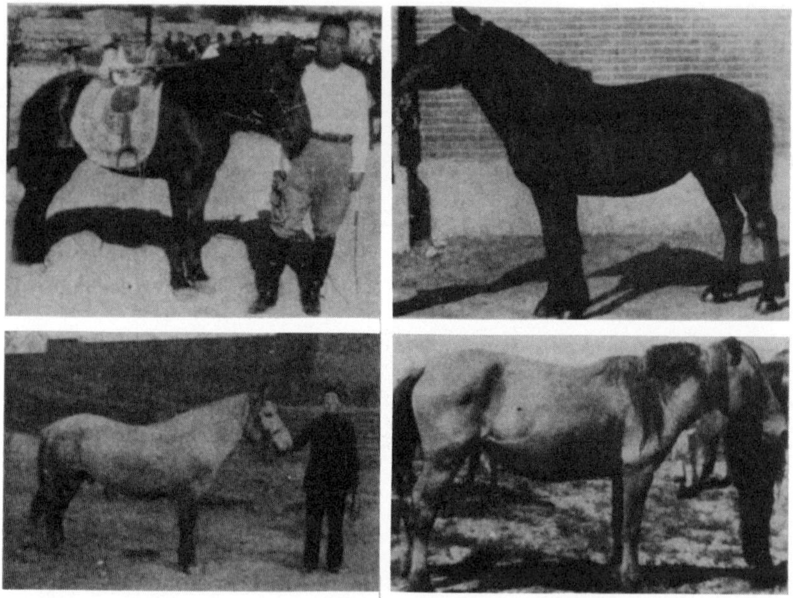

FIGURE 13.—Mongolian horses. *Upper left:* A Mongolian pacing horse of the best quality, shown with the owner, Prince Ta Li Tza Ya, of Alasan banner in Ningsia. *Upper right:* An exceptionally well-muscled horse. (Photograph by P. C. Tsui, Chinese Ministry of War.) *Lower left:* Stallion of a heavy type found in southwestern Kansu, near the Chinghai border. (Photograph by F. P. Sha, Chinese Ministry of Agriculture.) Horses of this heavy build are not numerous and are probably closely allied to the Sining horses. *Lower right:* Mare of eastern Mongolia, showing the rather common buckskin color.

The Sanpeitze and Hailar types are said to be similar to the Yilee. The Sanpeitze type is apparently found in both Outer Mongolia and Manchuria, since Army horse authorities give southeastern Mongolia as its native home, and the Japan-Manchukuo Yearbook (Wang, 1942) indicates that they are found in the western portion of Manchuria.

Many Mongolian horses are natural pacers and this gait has been encouraged by training and selection. The popularity of this gait apparently traces back to the days when mounted archers were used in warfare. When a pacer is traveling fast its body remains relatively level, so that it was possible to shoot with greater accuracy when riding rapidly past a target on a pacer than on a trotting or galloping horse. This use of the pacer no longer exists, and it is doubtful if the emphasis that is still placed on pacers in parts of the Mongolian horse area can be justified in the development of utility horses. The fast pace can be held by most of these animals for only a few hundred yards, and for utility riding the pace is not as satisfactory as a fast walk, a running walk, or a fast, easy trot.

Observations indicate that some of the good fast pacers actually do a rack rather than a pace.

South China ponies are found to some extent in the various provinces of southern China, but horses are not used as extensively here as in the northern and northwestern provinces. The ponies in various provinces are often referred to by the provincial name, such as Szechwan ponies, Kweichow ponies, etc. There is much variation in size and type within each province but the general size and type appears to be sufficiently uniform from province to province to justify including them all in one class. These animals average about 115 cm. in height at the withers. In general, these animals are better muscled than Mongolian horses, and the neck is more fully developed and the head is carried somewhat higher. Common faults of conformation are straight shoulders, low withers, steep croups, small leg bones, crooked hind legs, and small hoofs. Hoofs are of good quality, and stamina is reasonably good. These animals are sure-footed and become very adept at running up and down stone steps that are so common in the Chungking area.

Sikong horses are similar in conformation to the Szechwan ponies but somewhat larger in size. Measurements obtained by Lu (1943) of the height of 23 animals in Sikong varied from 111 to 130 cm. and averaged 121 cm. Weaknesses in conformation are the same as

FIGURE 14.—Sining horses. *Top left and right:* Two horses of the Sining type, that were in use on Governor Ma Hung Kwei's farm in Ningsia. *Lower left and right:* Two eastern Chinghai horses from near Datung, north of Sining. The saddle at the lower left is an elaborate version of the usual Chinese saddle, and that at the lower right is of much improved design and is called a Sinkiang saddle.

Figure 15.—South China ponies. ***Upper left and right:*** **Ponies typical of those found in Kweichow. The saddles are of foreign design and are not commonly used in China.** ***Lower left:*** **A pony of Kweichow hitched to one of the small carts used by the Ministry of Communications to supplement bus transportation.** ***Lower right:*** **A Szechwan pony hitched to one of the more elaborate modern carts used by the Ministry of Communications in Chungking and vicinity.**

those noted for other types, and a considerable number of animals are reported to have backs which are slightly high in the middle ("hog-backed"), and this is conducive to saddle sores. These animals are noted for their stamina and ability to climb, especially at high altitudes. Sikong horses are quite variable in size. Because of their geographic location it is probable that they are the result of interbreeding of Sining horses and South China ponies.

Owing to the nature of the horse and the types of tasks that it performs, actual performance data are difficult to obtain, but some figures are available that indicate the quantity of work done. Loads that are accepted as standard by the Animal Transportation Bureau are as follows:

Type of Horse	Pack		Draft	
	Weight	Distance	Weight	Distance
	Kilos [a]	*Kilometers* [b]	*Kilos*	*Kilometers*
Szechwan	60–80	40–60	600–800	40–60
Mongolian, Yilee, and Sining .	80–100	40–60	800–1, 000	40–60

[a] Kilo=2.2 pounds. [b] Kilometer=0.621 mile.

South China ponies are said to be able to carry 100 kilos and travel 30 to 37 kilometers in a day. The smaller ponies are said to be able to carry 50 to 60 kilos and travel 30 to 40 kilometers per day. Saddle horses of the Mongolian type can cover 40 kilometers per day on long trips and 55 kilometers or more on short trips. Mail carriers on the Kalgan-Urga route in the early days are reported to have averaged about 140 kilometers per day, traveling in relays. Commercial carts on the highways are usually loaded with 1½ to 1¾ tons. These loads are pulled by two or three horses and/or mules on the level and by four or five up the steep hills.

FIGURE 16.—Sikong horses. They are small and light-muscled but have good endurance, especially at high altitudes. Photograph at right was taken by T. Y. Hsu, Chinese Ministry of Agriculture.

Management and Feeding

In all parts of China the horse performs many tasks. It is used for riding, for packing, for farm and commercial draft power, and in the Army for cavalry and artillery work. Various sections of the country differ in the extent to which horses are used for the different purposes. For example, they are used to a greater extent for riding in the grasslands than in the farming areas. In some areas the "mule sedan" is still used for human transportation. Some mares are also used for production of milk. Wealthier Mongols may keep a number of mares to produce milk from which *arack*, a type of wine, is fermented. Mare's milk is also used to a limited extent as food. For example, house to house delivery with the mare being milked at each stop is practiced to some extent in Kiangsu. Such milk is valued especially for infants and invalids. Horse meat is also used to a limited extent in some areas.

In grassland areas very little or no shelter is provided for horses. In farm areas the typical barn has stalls, with manger, and is large enough to house all the farmer's workstock, which may include horses, donkeys,

FIGURE 17.—Horses and their relatives are used in many ways, especially in the dry-land farming areas of central and northern China. *Upper left:* Peking carts, in common use throughout North China for carrying people and light articles. *Upper right:* High-wheeled cart of a type commonly found in North China; photographed in Lanchow, Kansu. *Second left:* Modern heavy cart used in highway transportation. These carts carry $1\frac{1}{2}$ to $1\frac{3}{4}$ tons and put a terrific strain on the wheel animal. *Second right:* Showing common harness arrangement for hitching to a cart, and a bitting method that is too commonly practiced. *Third left:* A "mule sedan", a means of conveyance that is still used to some extent in North China. *Third right:* A horse and a mule pull a seeder and a packer in Shansi. In some of the drier loess areas, packers must be used to insure germination of seed. *Lower left:* Every farm has its threshing floor. Here a roller is being used to shatter out the grain on a farm in Shansi. *Lower right:* Gathering the wheat harvest on a farm in Shansi. Note the type of harness and mode of hitching.

FIGURE 18.—Tibetans use horses and their relatives extensively for riding. ***Upper left:*** **Servitors of a Tibetan regiment in Lhasa.** ***Right:*** **Tibetan cavalryman.** ***Lower left:*** **Tibetan merchants riding across the Tuna plain.**

mules, and/or cattle. Soil, which is brought in from the field and returned there after use, is used for bedding. Straw is used to a very limited extent as bedding. In most cases it would be eaten. Animals that are used in commercial transportation are usually housed at night in "mule-cart hotels" along the way. These are usually merely enclosures into which the carts are driven and the animals unhitched and tied, three, four, or five to one low stake. One side of the enclosure usually has a covered shed. Sanitation is poor and there is every opportunity for spread of disease and parasites. Separate hotels are maintained that cater to pack trains.

Harness types are quite simple in construction. Collars are made with a cloth face that is rough and conducive to collar sores. They are short and choke the animal when it is pulling. Also, the top is so made that the neck is apt to be rubbed and a sore or fistula develop. Hames are straight or nearly straight pieces of wood that are tied together at top and bottom. Traces are fastened to each hame at two points so that the pull is at or just below the middle of the shoulder slope. Traces are usually made of rope and are supported by another rope or strap over the back. The wheel horse in a cart has a heavy pad over the back which supports the front end of the shafts, and this animal is hitched directly to the shafts by short traces. The wheel animal also has a breaking strap (britching) which extends from the front of the shafts, around the quarters, and is supported by straps over the croup and extending forward to the shaft support. Farm horses may be hitched singly, in tandem, or side by side in a team with another horse or other draft animal. Hitches are made direct to a solid attachment on the implement, and the principle of the single tree and double tree is not used except in some instances by the Ministry of War. In some sections of the northwest the plows are made with long beams that extend up along the side of the horse nearly to the level of the back. This beam is attached to the hames by short traces, one direct and the other leading over the back to the opposite side, the back being protected with a pad.

Horses in carts may be hitched singly or in tandem. If the load is heavy a third animal may be hitched beside the front animal, with traces passing through a ring at the front of the shaft and thence to

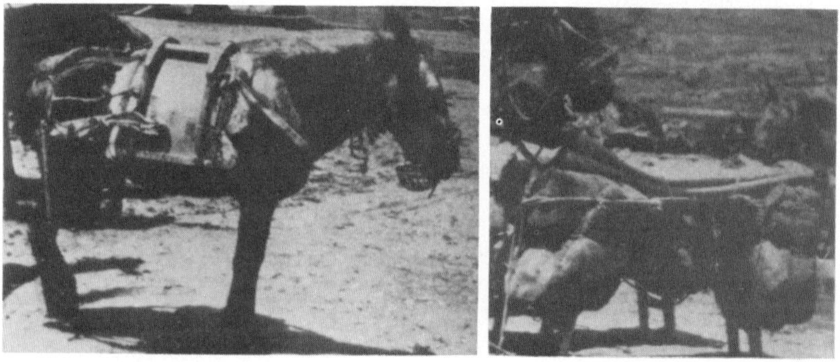

FIGURE 19.—The Chinese packsaddle. *Left:* A pad is placed on the back and the base of the saddle is placed over it. *Right:* The frame that holds the load. This is loaded, then lifted on the animal, and can be removed whenever necessary without disturbing the packed load.

the cart. This animal cannot work efficiently owing to the fact that it pulls at an angle rather than in a straight line, and it also reduces the efficiency of the other two since it puts a side pull on their efforts. The ropes also wear sores on the quarters of the horses that are hitched at the side. The wheel horse is subjected to considerable punishment, since at every movement or lurch of the cart the full weight is thrown on it. When cart trains reach a hill, the practice is to leave half the carts, using four or more animals to pull the other half up the hill, then return for the second half.

The small Peking cart, which could be handled easily by one horse, was a very useful vehicle, but this has been expanded to carry loads of 1½ tons or more in commercial transportation, and the design is not suited for such loads. Drivers attempt to compensate for the inadequacy of design by putting their strongest animals at the wheel, but

FIGURE 20.—*Top:* Scene in the horse section of a livestock market in Shansi. *Bottom:* One portion of a "mule-cart hotel" or caravansary. These are found along the caravan routes throughout the northern portion of China. It is unsafe to stop outside these or other walls for the night. They are excellent centers for distribution of disease and parasites.

the loads and the strain are too great for an animal to bear without damage.

Packsaddles consist of three parts: A pad to protect the back; a framework to support the pack, which has a strap extending forward and under the neck and a second strap or crupper extending around the quarters or under the tail, and with a support over the croup; and the pack frame, which is loaded and then placed on the animal's back.

615999—45——3

Cruppers for use under the tail may be made of leather or of skin from wild-horse (*Equus hemionis kiang* and *typicus*) legs, and also of wood. Cruppers for use around the quarters may be of leather or a stick with straps or ropes attached.

Saddles for human use are made with a basic framework similar to the packsaddle base. Two boards rest on either side of the backbone with a space between so that the backbone itself is not rubbed. On this frame is built a high seat that slopes up to the front and rear, being symmetrical or nearly so when viewed from the side. These saddles are not comfortable to sit in, and stirrup straps are usually rough and not protected with shields, so that the legs are apt to be rubbed. When riding for any distance a blanket or quilt is folded and tied on to the saddle. Some persons claim that a horse of 15 hands (152 cm.) is too tall for practical use by Chinese, yet it is customary to put one of these high saddles, plus three or four inches of padding, on a 13–2-hand (138 cm.) animal, making it just as high and more difficult to mount than a 15-hand animal with a low saddle. There is a type of saddle in Sinkiang that is so shaped that it is a comfortable seat and is built low to the back. This saddle is similar to the low-type American and English saddles, except that it is covered with cloth instead of leather and has a small horn.

Bridles are of simple construction. Bits may be either straight or hinged in the middle. Wooden hackamores are used in some areas. Reins may be of rope, hair, or leather and are usually rough and apt to blister the hands of the rider.

Hobbles are in common use in the grassland areas and consist of ropes with convenient loops for placing around the lower cannon or pastern. Single animals may be hobbled by tying the front legs or one front and one rear leg together. Two animals may be hobbled together by tying their left front or right front legs together, or the left front leg of one horse may be tied to the right front leg of another. Another scheme, used in Ningsia, is to tie the front legs of an animal together and connect this hobble with one of the rear legs.

Little information is available on the methods of breaking and training of horses. Two methods are said to be in use. One is a "slow" method by which the animal is gradually accustomed to being handled, first with a halter, and later with a bridle if it is to be ridden or driven with lines. Then it is harnessed or saddled and gradually accustomed to work. The other is a so-called "fast" method in which the horse is tied to a post and whipped for an hour or so. When its spirit is completely broken it is hitched or saddled and put to work. Lack of adequate training is most obvious in saddle horses. Many of these animals have tough mouths and are therefore very difficult to handle. The average animal is accustomed to going its own gait and when allowed to break out of a walk is apt to grab the bit in its mouth and run. Many Tibetans, Mongolians, and others who live in the open grasslands depend to a large extent on horses for transportation and their animals are often well broken and trained for the jobs they expect them to do.

Draft and pack animals are usually driven by voice or with a whip. Cart drivers in commercial and other highway transportation walk beside their animals, directing them with voice and whip and occasionally grabbing the traces to pull them to the side of the road. Lines are used to guide the animals in a few instances, such as the small taxi carts that are used for human transportation in and near Chungking. Farmers drive almost entirely by voice when plowing or doing other farm work. With pack trains, the driver usually walks behind, exercising only such control as is possible with his voice.

In Mongolia a piece of red cloth may be tied to the tail to indicate a kicker, or to the forelock to indicate a biter or striker.

Horses that are used in commercial transportation, either in carts or as pack animals, are usually overloaded. In fact, an overload is the rule in China, not only for draft animals but for cars, trucks, and coolies as well. Because of this overload, and also owing to inadequate nutrition and other causes, many horses in China are no longer useful at 10 years of age and a horse of 15 is considered quite old.

In some areas horses are tied with the head held high so they cannot lie down, owing to a belief that a horse only lies down when ill. The practice more probably has its actual origin in the use of this technique to prevent the animal's getting soiled, thus avoiding the necessity of grooming.

A high feed trough is used in some sections of Kweichow with the belief that this will cause the horse to have more arch of neck and therefore appear more spirited. In Shansi many carters and farmers take animals out to roll after a day's work, and when working hard they may be fed in the middle of the night.

In general, the art of good horsemanship, as it is known in western countries, is almost unknown in China. However, there are many, particularly among the nomadic groups, that know horses well and handle them efficiently as judged by their standards and the conditions under which they live and work.

In the grassland areas grass is the most important feed for horses. It is reasonably adequate during the warmer months, and in many areas the horses become quite fat during this period. During the winter months the dry grass does not give sufficient nutrients and the animals become thin and their resistance is lowered. Losses are frequently heavy during bad storms when snow is too deep for the animals to find feed. Grass may be cut and carried to horses where it is available in the farm areas. The heaviest snowstorms and resulting losses usually come in March, April, and May in many parts of the grasslands.

When herded on grassland, horses are kept on the high mountains from late June through July and August to avoid heat and flies. During September, October, and November they are kept on the lower slopes and during the remainder of the year in the valley areas.

Where grass is not available, straw or stalks (rice, oat, millet, etc.) are the only roughages fed in most cases. Straw is usually chopped into short lengths before feeding. This is as short as ¾ to 1 inch in Shansi and Hopeh, while in other areas the length is approximately 2 inches. Iris and other non-poisonous plants of low palatability are cut in the fall for roughage in Ningsia.

Bran (wheat), peas, kaoliang, and black soybeans are the grains commonly fed to horses, the grain most commonly used depending upon the type available in each area. Bran is usually given by dampening the straw and mixing the bran with it. Soybeans are often soaked before feeding. Peas and corn, in areas where they are fed, are usually given whole.

Straw is usually dampened when fed alone. In some areas horses are watered before feeding and in others, after feeding.

Salt is not given to horses with any regularity or in sufficient quantities. Animals are frequently seen licking up urine and even feces and dirt in their search for salt. Salt is scarce and very expensive in many areas, so that many of the people cannot afford to buy salt, either for their animals or themselves.

Night blindness is reported to be common in Kweichow and adjacent regions, indicating a possible deficiency of vitamin A.

In general, horses are produced in the more remote areas such as on mountain ranges and in the grasslands of the west and northwest. From these areas the excess males move into the more heavily cultivated areas and into commercial use. Hence, there is a preponderance of females in the remote areas and of males in areas where most horses are used for draft purposes.

Most animals are apparently sold at private treaty and at the large livestock markets that are held occasionally in certain areas. Tea bricks are frequently used for purchase instead of cash in the more remote areas.

By-products of the horse are obviously few. The hides of dead or injured animals are usually salvaged for leather and in some areas bones may be gathered up and sold for use in fertilizer. Hunts for wild horses are sometimes organized to obtain hides for leather and skins of legs for cruppers.

Breeding

No data are available on the age of puberty, but there are apparent differences in the rate of sexual development. Stallions of the Sining type are said to be ready for active service when 2 to 3 years of age, while Mongolian stallions from Ningsia and Suiyuan are said to be not ready for service until 3 or 4 years old. Where hand-breeding is practiced, mares are usually bred for the first time at 2 or 3 years of age, but in range herds they are more apt to be bred at whatever age they first come in estrus.

Very few data are available on the duration of estrus and length of the estrous cycle. These are estimated to be 2 to 5 days and 18 to 25 days, respectively, in different areas.

The percentage of pregnancies obtained during a breeding season is variously estimated at 20 to 80 percent. It is probable that the average for the country is below 50 percent.

The breeding season extends from February to October, with the bulk of the breeding being done in March, April, May, and June. Records obtained by Sha (1943) at the First Horse Breeding Station, near Wukung in Shensi, show the following distribution of breedings of farmers' mares that were brought to the station for service in 1942:

Month	*Feb.*	*Mar.*	*Apr.*	*May*	*June*	*July*	*Aug.*	*Sept.*	*Oct.*
Number	20	152	170	70	42	43	13	3	2

In Sikong, it is reported that most mares are bred during June and July. These are the first months when the supply of green grass is good after the winter season.

Accurate data on length of pregnancy are not available but it is about 11 months, according to most observers.

In most areas the foal is allowed to run with the mare until the mare weans it. In some Tibetan areas such as the Lihua plateau, foals are blanketed for a month or so after birth to protect them from abrupt temperature changes that are common at this high altitude.

When hand-breeding is practiced, mares may be hobbled with a rope which goes around the rear legs above the hocks, passes forward either between or to one side of the front legs, and is tied around the neck. This provides some restraint, but a vigorous mare can kick a stallion if she resents his advances.

In some areas, southern Shensi for example, the farmer insists on having the cervix of the mare "opened up" by hand before service. This practice is also common in other countries, but there is no evidence that it serves any purpose other than to provide an extra chance for introducing infection.

In Kweichow some farmers make a practice of breeding their mares three days after foaling, whether in estrus or not. Evidence in other countries indicates that foal estrus occurs about nine days after foaling, on the average, but there is considerable variation. If foal heat occurs earlier in certain Chinese horses, this should be definitely determined. In any event, it is sounder practice to delay breeding until the first estrus following the foal estrus, thus giving the uterus time to clean out and be in better condition to begin another pregnancy.

Many males are left intact even though they are not to be used for breeding. This is especially true in South China where males of the South China pony type are seldom gelded, but the practice is prevalent in other regions. Most males that are not used for breeding at the Ministry of War's horse stations are gelded.

In the grasslands where horses are run in herds, the usual practice is to supply 1 stallion for every 10 mares. In Sikong and other grassland provinces, most of the good stallions are owned by the ruling class or by the lamaseries.

The French apparently brought 100 stallions (probably of Arab or Barb extraction) to the northeast in 1900, from North Africa. These were dispersed in 1907 and sold in Inner Mongolia and adjacent areas. Since no organized breeding program was followed, their influence was diffused and absorbed into the native blood, where traces of their influence can still be seen.

Russian horses of uncertain origin have been used in the border areas of Sinkiang, Outer Mongolia, and Manchuria and have brought about some increase in size. No definite breeding plan has been followed, and no clear-cut information is available on the results obtained. Fifty Russian stallions (probably Orloff Trotter and extractions from this breed) were brought into Sinkiang in 1937, and it is reported that their progeny have greater size and strength than the native horses.

Arabians have been introduced and used to some extent in the Ministry of War's horse-breeding program, but no results of this breeding work have been published. Some work has also been done with grade thoroughbreds from Australia, but no results have been published.

Some foreign stock, especially some animals for racing, was introduced into coastal cities, but there is no evidence that these animals were used in any planned grading-up program, and they apparently had little effect on the quality of horses produced in those areas.

According to the Manchurian Yearbook, the Japanese were using or planning to use Arab, Anglo-Norman, Anglo-Arab, and Hackney horses for some grading-up operations in Manchuria.

Donkeys

Historically, the donkey was developed and used by the Chinese farmer while the nomads used horses, and this relationship still holds to a considerable extent. As the farmer moved into the edge of the nomad grasslands, he took donkeys with him, and horses gradually filtered into the farming areas. The donkey is probably the most important draft animal in North China.

Figure 21.—Distribution of donkeys in China. *A*, Approximate area in which donkeys are numerous and important in agriculture and transportation. *B*, Areas in which donkeys are of relatively little importance. *C*, Area in which a considerable number of large-type jackstock is found. The distribution of mules is similar to that of donkeys.

Types

Chinese donkeys probably vary more in size than in any other important characteristic. Mature animals vary in height from approximately 90 to 137 cm. For convenience, they may be divided into three types: large, intermediate, and small. There is no clear-cut line of demarcation between these; one type blends into another. The legs are apparently shorter in relation to height in the smaller types.

The large-type donkeys are found primarily in the Wei and lower Yellow River Valleys. Some of these are also found in Manchuria,

FIGURE 22.—Chinese donkeys vary considerably in size. *Upper left:* Large-type jack of Shensi. *Upper right:* Smaller-type jack of Shensi, height 40 inches. *Lower left:* Small-type donkey, height 37 inches. This one was photographed in Szechwan but such small ones are found in all parts of China where donkeys are used. This one was used by a quack doctor. Medicines are carried in the pouches, the patient is seated on the donkey for diagnosis, and the paper "wings" advise prospective patients to get well quick on a horse. *Lower right:* A donkey pack train carries salt from Ningsia into Kansu.

where they have apparently been taken by emigrants from Shantung and adjacent areas, and intermediate ones are also found in that area. The small type is the predominating type in Sikong but is also found to some extent in almost all areas where donkeys are used. The number of donkeys in Sikong is relatively small. The donkeys in Ningsia also are predominantly of the small type. The intermediate type appears to predominate in many of the other areas where donkeys are found in large numbers.

There is wide variation in the quality of animals within the three size groups. Some have reasonably well-developed necks, with the head carried fairly high, medium development of withers, straight backs, level croups, reasonably deep bodies, and fairly straight legs. Others have some or most of the faults that are the opposite of the more desirable characteristics listed above. Crooked rear legs are especially common. Colors usually seen include buckskin (various shades of dun with dark stripe down the back and with dark mane and tail) and various shades of brown, some being almost black, with lighter color on the underside. Black and white are seen occasionally. Albinos are rare.

Data on performance are very scanty. Pack trains usually cover 35 to 40 kilometers per day, but the better animals may cover as many as 55 kilometers. Loads for animals of the intermediate type usually vary from 80 to 100 kilos.

Some donkeys have short, smooth coats, while others have long and rather rough coats. Others are intermediate. Long coats are most common in the small type.

Management and Feeding

The large-type donkeys are used primarily for draft purposes on farms and in commercial transportation and also as sires in mule production. The intermediate and small types are both used for farm work, for riding, and for packing.

Housing, harness and other equipment, and methods of management are very similar to those described for horses. It is generally conceded that the donkey can exist with less feed and care than the horse, and this reputed ability is frequently taken advantage of by owners.

One interesting practice that is different from that used with horses is the habit of sitting on the hips of the donkey when riding, so that the rider feels the movements of only the rear legs.

Feeding practices are essentially the same as those described for horses. In southern China some farmers give their jacks two eggs for each time they are allowed service.

Breeding

Breeding practices are similar to those described for horses. The donkeys differ in having a longer pregnancy, which is estimated to be about 12 months. Estrus is also said to be easier to detect in Chinese jennets than in mares.

Many jacks apparently breed either mares or jennets with equal desire. Jacks that are accustomed to serving jennets are frequently hesitant about serving mares, but little difficulty is reported where the jack is used first for serving mares.

Mules

The distribution of mules in China is similar to that of donkeys.

Types

There are no clear-cut types, the most important variation being that of size. The mares that produce them vary somewhat in size

FIGURE 23.—The mules produced by mating the large-type jacks to Mongolian mares are very useful animals, highly prized for heavy work. *Upper left:* A buckskin color frequently is seen. *Upper right:* Young mule in Ningsia, broken to the saddle. Many of these mules make good riding animals. *Lower left:* A 13–3–hand mule carries the senior author (R. W. P.) into Tibetan country around Kokonor. *Lower right:* A Shansi donkey turns one of the pumps of a type that is commonly used for obtaining irrigation water in Shensi, Shansi, and some other sections of North China.

and type, and the jacks are extremely variable in size. Some of the larger, stronger mules stand as high as 137 to 142 cm. and have reasonably good conformation. Manchurian mules are reported in the Manchurian Yearbook to vary from 1.3 to 1.6 meters in height.

Data obtained on mules in use on the highway at Yungchang, Szechwan, in the summer of 1943, indicate the size of animals in use in that area. These data, supplied by Hsu (1943), are summarized below:

Item	*Males*	*Females*	*Geldings*
Number of animals	35	18	2
Length, top of wither to base of tail . . . cm	105.7	108.0	119.3
Height at withers	126.3	115.7	134.0
Heart girth	141.0	144.7	157.3
Middle girth, in front of last rib	160.0	163.0	171.3

The estimated age of the males ranged from 2 to 10 years and up, and that of the females from 2 to 10. The two geldings were 6 and 15 years of age. The chief deficiencies observed in mules are that they are frequently too light in body and bone for draft purposes; and short straight shoulders, sloping croups, light muscling, and crooked rear legs are also frequently observed.

The major portion of the large mules is produced in the Wei and lower Yellow River Valleys. Shensi and Shantung are both noted for the production of good draft mules. Many mules are also produced in Suiyuan and Manchuria. Mules are also produced in other areas, but these have become noted as the more important centers of production.

Data on pulling power and other items of performance are not available. In packing, it is customary to load mules with 10 to 20 kilos more than is placed on horses of equal size. In commercial transportation, the larger, stronger mules are usually used as wheel animals in the heavy carts. These carts are usually loaded with 1½ to 1¾ tons and the wheel animal bears the major portion of the punishment in handling the load. Some of the larger mules are quite satisfactory for riding purposes and are especially useful in rough country. Many mules are good walkers and some have a very satisfactory running walk. This type of gait is seen frequently in Ningsia and is also found in other areas. In marshy country, mules are not as satisfactory as horses because they are more apt to bog down, owing to their smaller feet.

Management and Feeding

The mule is used in commercial cart transportation, in packing, for farm power, for pulling light carts, and for riding. The larger mules are much sought after for commercial transportation, but some are used on farms and for other purposes. Medium-sized animals are used for all purposes, while the smaller ones find their greatest usefulness on the smaller farms and for packing.

Housing, feeding, and other practices are similar to those described for horses.

Breeding

Mules do not reproduce, except in rare cases; hence there are no breeding problems with mules themselves.

The practice of leaving many males intact is similar to that for horses and donkeys, owing to the belief that the intact male is a better work animal than the gelded one.

Certain provinces have reputations as mule-production centers, and the source of mares differs in the different areas. Breeders in Shantung buy many mares from Inner Mongolia for this purpose. In Shensi some of the mares are produced at home while others are purchased from Kansu and some from Inner Mongolia.

In mule-producing areas, the service charge for jacks may be higher than that for stallions. In and around Wukung, Shensi, the service charge for the large jacks is about three bushels of grain. The farmers in this area are anxious to breed mares to jacks for mule production, and in order to encourage the breeding for horse production, the First Horse Breeding Station makes no service charge for stallions and one third the regular local rate for jacks.

Improvement of the mule is dependent upon improvement of horses and donkeys. Discussions of the need for improving the horse for work and for providing mares to produce mules are frequently heard, but the need for improvement of donkeys is usually overlooked, the lowly donkey being merely taken for granted.

Camels

The area in which camels are produced and used is outlined on the map in figure 24. It includes Sinkiang, Outer Mongolia, most of Inner Mongolia, and portions of Jehol, Kansu, Chinghai, and the edge of Tibet. Caravans travel outside this area and occasionally go as far south as Chengtu, Szechwan.

FIGURE 24.—Distribution of camels in China. ***A*****, Area where camels are produced and used.** ***B*****, Area where almost no camels are produced and relatively few are used. Many camels travel outside of** ***A*** **in commercial transportation to Lanchow, Kansu, and to Peking, Hopeh, and vicinity.**

Types

Practically all the camels in China are of the two-humped type. The single-humped type, or dromedary, is said to exist in Sinkiang, where it has probably been brought in the normal course of commerce with regions to the west. It is reported that there are a few single-humped camels in Manchuria.

There are no distinct types or breeds, and there has been no attempt to develop or isolate any specific type or types. In general, compact, low-set, massive, well-muscled animals are desired. Colors range from dark brown to very light dun or cream. Some are nearly white.

FIGURE 25.—*Top:* Camels grazing on camel thorn and other desert vegetation in Ningsia. *Bottom:* A choice breeding bull in the foreground stands beside a bush of camel thorn. Camels bite off this very thorny bush by the mouthful and nonchalantly chew it up.

A purplish red color is said to be most desired in some areas. A white blaze or spot on the forehead or face is unpopular since it is supposed to bring bad luck.

No weights or measurements are available on Chinese camels, but mature animals are said to measure from about 150 to 210 cm. at the top of the hump. One large camel that was slaughtered is reported to have yielded 235 kilos of meat.

Accurate data on working capacity are likewise unavailable. Camels are reported to travel 35 to 40 kilometers per day on long trips and up to 85 kilometers on short trips in 24 hours. The strongest camels carry 275 kilos and cover 1,150 kilometers in 30 days. Camels of average size and strength may carry about 140 kilos at the beginning of the working season but only 120 at the end. In some areas such as Ningsia the average load is reported to be about 115 to 140 kilos. Cross (1917) gives some figures on the loads for camels in India that are interesting for comparison with the previous figures obtained from camel-owners in China. He states that the load allowed on government camels is 182 kilos, but that natives load as heavily as 270 to 360 kilos. He reports that the strongest camels can carry up to 435 kilos. These loads are apparently for dromedaries. Cross also states that camels can walk at the rate of 2½ miles (about 4.2 kilometers) per hour and if at continuous work should not be allowed to cover more than 20 to 25 kilometers per day, and he recommends one day of rest in each week. Camels in Ningsia are reported to travel about 2 miles an hour with packs.

Cross also states that a good riding camel can travel at the rate of 6 to 7 miles (about 10 to 12 kilometers) per hour and cover up to 100 or 115 kilometers in a stretch. If at continual work they can average 50 kilometers per day for several days.

Camel's hair or wool is used for making blankets, certain clothing materials, and other textiles. It is fine in the young animal and of relatively poor quality from old animals. The price in northwest China is reported by Chang (1941) to be three times that of sheep's wool. The average yield is said to be about 2.4 kilos, but large animals yield as much as 3.5 kilos. The production of camel's wool in Manchuria was reported (Wang, 1942) to be 71,419 pounds in 1937. The estimated number of camels was 12,800. If these figures are accurate, then the yield per head was about 5.6 pounds, or slightly over 2 kilos. Animals are seldom shorn, since shearing results in too-sudden change and camels may become ill. Shedding begins in April and is complete by about the end of May. The wool falls out in bunches and is picked up, or it may be pulled out when it loosens. The stealing of wool by children, by pulling it out when caravan camels are resting, is common in some areas. Camel's wool is soft but is uneven in fineness and is apparently difficult to spin since most of the yarn seen on the market is uneven and bunchy.

The milk of the camel is similar to cow's milk. Chang (1941) gives the following comparative figures:

Constituent	*Camel (Percent)*	*Cattle (Percent)*
Fat	2.90	3.69
Protein	3.67	3.55
Lactose	5.78	4.88
Minerals	0.66	0.71
Water	86.94	87.17

Production figures are not available on Chinese camels, but milking females are frequently taken on trips to supply milk for the travelers. Cross states that Indian camels give as much as 1 to 2 gallons at 1 or 2 weeks after parturition and continue lactation for 8 months under desert conditions and up to 18 months under conditions of good feed.

The skin of camels is used to some extent for leather. Camel manure serves as fuel on the desert after it dries. Camel meat is also eaten in some areas but is not palatable to many people. It is said to be rather coarse and dark in color.

Management and Feeding

The primary use of the camel is for transport of goods and for riding in semi-desert and desert areas. Wool, milk, skin, meat, etc., are merely by-products of this main function.

Camels usually start to work at about 4 years of age. They may be used for riding from 3 or 4 years to 5 or 6 years and then put to work in caravans. They are reported to be used for work until 16 to 22 years of age and may live to about 30 years. Cross states that Indian camels may be put at light work at 4 years, but that some poorer owners start working them at 2 or 3 years. He believes that they should not be worked until 5 years of age. He states that a camel is at its prime at 8 years and may be worked for a total of 6 to 19 years and longer if conditions of feed and management are good. Leonard (1894) claims that camels should not be worked until 6 or 7 years of age.

The training process is simple. A bamboo ring or a straight stick with a knob on one end and with the lead rope fastened at the other end is placed in the nose of young camels. They soon become accustomed to it and are easily led, following older camels. They must be accustomed to loads very slowly and with only light loads at first, else they become balky. They often object to being loaded and some develop a habit of kicking with the front feet (striking), particularly the males. Such animals may have a red rag tied on the forelock to identify them. Some have vile tempers. Males are apt to be dangerous during the breeding season, and it is often necessary to muzzle them at this time.

One man handles seven or eight camels in caravans. He leads the front animal and the others are tied in tandem from the nose ring to the pack in front. Only a very small rope is needed if the animals are gentle.

Shoe leather may be sewed to the pads of the feet to prevent injury to the feet when traveling or to protect feet that have worn thin from travel.

Packsaddles like those used on horses and cattle are not used on camels. Instead the packsaddle consists of two bags, made of wool and filled with straw, which are placed along either side of the animal's back. Either one or two wooden bars or poles are fastened to each bag, running along the side of the body. The bags, with poles attached, are held in place by ropes which pass around the body. A piece of felt may be placed over this equipment. The load is divided into two equal parts and one half is attached to the pole or poles on each side. The load may be in sacks or in the form of bales depending on the nature of the material being carried.

In summer camels live on grass in the mountains. This allows them to fatten so they are in condition to work during the remainder of the year. Humps become thin and fall over during the winter but fatten and become erect during the summer grazing season. Grazing in the higher areas also enables them to miss the worst of the summer heat. While working they get some grass by grazing during the day. Other roughage includes wheat, rice, oats, and millet straws. The straw is chopped into about two-inch lengths. Peas are the only commonly used concentrate. These are cracked (coarse) and soaked in water before feeding.

Camels require considerable salt, which is usually spread on grass for them to pick up. No salt is given while on summer range.

Water is given once daily, in the morning. It is said that camels can go without water for three days without suffering because "they store water in their humps". There is no direct evidence for this popular belief. One study indicates that the camel is probably able to go for long periods without water because of the relatively large amount of water produced as an end-product of metabolic processes in the body. Wang (1941) states that the hump is composed of a connective tissue network and adipose tissue. Wang describes the digestive organs of the camel and points out that the rumen has two "water bags" or out-pocketings which are not found in other ruminants. He mentions that the contents of these bags are moist or liquid and similar to the contents of the reticulum. It seems probable that the name "water bag"

615999—45——4

may be an improper designation which indicates a supposed function that may not exist.

Camels are frequently worked at night, in order to avoid the daytime heat.

The camel has 34 permanent teeth, but writers differ on the classification of these. Wang (1941) lists 2 incisors, 2 canines, 6 pre-molars, and 6 molars in the upper jaw; and 6 incisors, 2 canines, 4 pre-molars, and 6 molars in the lower jaw. Cross lists 6 canines and 10 molars in the upper; and 6 incisors, 4 canines, and 8 molars in the lower jaw.

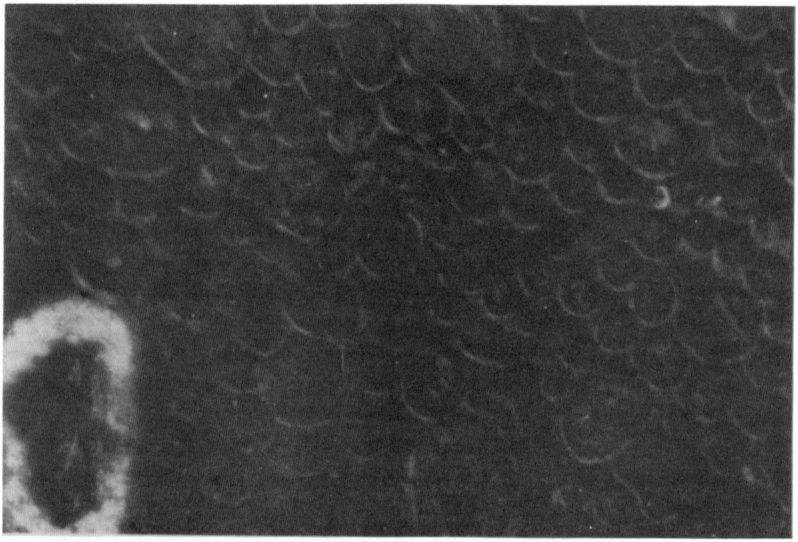

FIGURE 26.—Cross section of camel's wool, magnification—370x. Note the extreme variation in size of fibers.

Cross (1917) also outlines the relation of teeth changes to age, as follows: 2 temporary incisors appear at 5 to 12 days, and by 1 year there are 6 temporary incisors. By 4 years these are worn to the roots. At 5 years there are 2 permanent incisors, at 6 years there are 4, and at 7 years there are 6. The first canines appear at 8 years, and after this, age changes can be detected to some extent by the amount of wear.

Some camel's wool is used by the local people for hand-spinning and hand-weaving. Before the war, most of the wool from Inner Mongolia that was not used locally went to Tientsin for export. Wool from Sinkiang probably goes primarily to the U.S.S.R. Some camel's wool comes to such centers as Sining and especially Lanchow, from where it is distributed to various manufacturers in the area; and some of it goes as far as Chungking.

No specialized marketing techniques have been developed. Animals are apparently sold for the most part at private treaty. The best animals in northern Kansu are valued at about $100 (silver) while average animals are valued at about $80. Wool goes through about the same channels as sheep's wool, and little attempt is made to grade it, to properly package it, or to sell it on a quality basis. In Ningsia camel's wool is sorted by hand into three grades or degrees of fineness.

Breeding

Data on breeding practices are very scanty. The breeding season is reported by some to be from August to October and by others to be as late as January. Clear-cut information on estrus, length of the estrous cycle, and duration of pregnancy under Chinese conditions is lacking. Pregnancy is estimated by different men to be from 12 to 17 months, but the most reliable estimates place it at 12 to 13 months. Young are apparently born in the late winter and early spring, in most areas. The females usually produce young every second year. One breeder in Kansu estimated that 80 percent of the females bred in a season produced young and that 50 percent of those bred had young alive at weaning. Reports from Ningsia indicate that the level of fertility is not as high as 80 percent, but the death losses are much less frequent than the foregoing figures indicate. Single births are the rule and twins are very rare.

Young camels are reported by one owner to be weaned at 3 months if the dam is working, but, if idle, weaning may be delayed until 6 months. It is said that the young graze readily by 2 months of age. Weaning at 8 months or later appears to be a more common practice, and in Ningsia females do little work and the common practice in some areas is to leave intact 1 male for each 10 females. The method of castration is similar to that in other animals, which is to make incisions in the scrotum with a knife, pull out the testes, and sever the cord with a knife. It was reported by one owner that this operation was done at 6 months but Wang states that the testes are still in the abdomen at 6 months of age.

Males usually run with the herd. Before copulation, the male forces the female to the ground by biting or pressing on the neck. He then mounts, but in a sitting position with the hocks on the ground. This position in mating is necessary apparently because of the structure of the reproductive tract and sheath, which opens backwards.

Cross (1917) gives some information on breeding practices with camels in India that are interesting for comparison. He states that the duration of the first pregnancy is usually about 12 months and 20 days, and that later pregnancies are usually just a few days over 12 months. Females usually produce young every second year but occasionally breed again about six months after parturition. He suggests December and early January as the best time to breed under Indian conditions. The percentage of infertility is reported to be very low. He states that females are usually bred first at 4 years, but if well fed will show estrus at 3 years, and will continue to produce until 20 years and sometimes beyond this age. He describes the vulva as being slightly swollen during estrus, and the female as being nervous and seeking the male. He reports that the act of copulation requires about 15 minutes, and that other camels gather around and protect the pair while they are mating. The process of parturition is described as rather difficult and he cites the necessity of tying the female when labor begins so she will not run and injure the young. The placenta usually comes with the young or within one-half hour afterward. Cross also states that the young should be weaned by 15 months or the next lactation will be short. He says that one male can breed 30 to 50 females, but that 30 is the best number. He recommends that males not be used for breeding before about 8 years of age. Castration is also recommended, but he points out that bone and body development will not be adequate if this is done before 6 years.

Leonard (1894) states that females suckle their calves for from 1 to 1½ years and do not come in estrus during this period. He states that one young is produced every three years by a female, and that one male can serve 15 to 50 females, but he recommends 16 as the optimum number. He also claims that females should not do any work during gestation or lactation or at all events for a year after birth. He claims that there is frequently great mortality among newly-born camels owing to working the dams during gestation, weaning the young too soon, and working the dams too soon thereafter.

Cattle

All cattle in China are commonly referred to as "yellow cattle", a term which distinguishes them from water buffalo and yak.

Types

The cattle of China may be divided into two main classes, the Mongolian cattle of the northern and northwestern regions and the draft

Figure 27.—Distribution of types of cattle in China. *A,* Area in which Mongolian cattle are found. *B,* Area in which South China draft cattle are found. *C,* Area in which a considerable portion of the cattle are of a large type. Some large-type cattle are also found in the Szechwan-Hunan border area. Some very small cattle are found in Sikong and Tibet, similar to hill cattle of northern India.

cattle of the farming areas of central and southern China. It is probable that a third classification is needed for the small cattle of the Tibetan plateau regions.

The Mongolian cattle are light-boned and light-muscled, being somewhat similar to the dairy type in general build. They are reputed to be somewhat better milk producers than the heavier draft animals but their production is not sufficiently high to characterize them as dairy animals. In color, they are quite variable. Some are black, some brown, some yellow, and there is some white spotting. Many brindle animals are also seen.

FIGURE 28.—Mongolian cattle of northern Shensi. (Photographed by F. P. Sha, Chinese Ministry of Agriculture.)

Kislovsky (1938) described the cattle of Outer Mongolia as being a distinct type from European cattle or the cattle of southern Asia. He described the type as non-specialized and primitive and weighing about 275 kilos on the average. Wide variations were observed in dimensions, color, and markings.

The draft animals of central and southern China are of a more muscular type than the Mongolian cattle and resemble dual-purpose cattle in general conformation. The most common color is yellow but there are some browns, and in some areas various shades of red are

FIGURE 29.—Mongolian cattle, native of Ningsia. These cattle are raised almost entirely for draft purposes, but in some areas Mongolian cattle are milked.

seen. Blacks are also seen but are not desired. Farmers have a saying, "Ten blacks, nine lazy". Another color is called "horse-wasp color", in which the forehead, shoulders, hump, and backline are yellowish brown and the remainder of the body is black. This is most distinct in the spring, and in the autumn these parts turn black or nearly so, except for the forehead and shoulders. The animals in the Wei and lower Yellow River Valleys are of a large type and some large-type animals are found in the southeastern portion of Szechwan and adjacent regions of Hupeh, Hunan, and Kweichow. The animals in other portions of the southern region are predominantly of a small type.

There are marked variations within the type found in any given area. In the Wei and lower Yellow River Valleys, for example, there

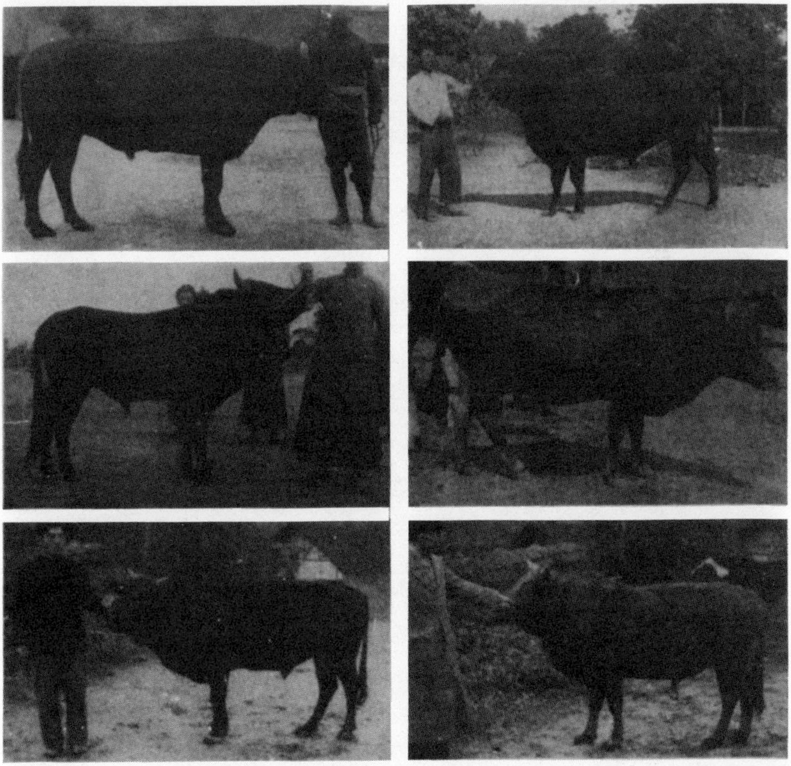

Figure 30.—Draft cattle of central and South China. *Upper left:* Draft bullock of a large type commonly found in Shensi. *Upper right:* An exceptionally fine bull in Shensi. *Middle left:* A first-prize bull at a local fair in Szechwan. (Photograph by T. Y. Hsu, Chinese Ministry of Agriculture). *Middle right:* Szechwan yellow cow. *Lower left and right:* Two bulls of the type commonly found in Kweichow.

are at least two distinct types. The animals in southern Shensi are large and well muscled and quite massive in conformation. The predominant color is a reddish brown. The animals in Honan are taller but are not as heavy since they are much thinner-bodied when viewed from front or rear. They are quite deep-bodied and have long legs, and the walking gait is quite rapid. They are lighter in color than the Wei Valley cattle, being more of a dark yellow tinged with brown.

The Mongolian cattle have no hump but most of the cattle in the south have a small hump on the top of the shoulder. One gets the impression that the size of the hump tends to decrease as one travels

FIGURE 31.—*Top:* A large-type bullock pulls a cart in a harvest field in Shansi. *Bottom:* A yoke of cattle of a small type typical of the southern slopes of the Tsingling Mountains, in Shensi, pulls a typical Chinese plow.

from the southern to the northern portion of this southern region, but this may be due partially to the custom of leaving more of the males intact in the south and castrating a somewhat greater portion of them farther north. There is considerable variation in the size of the hump in any given area.

The following measurements indicate something of the size of cattle in Hunan, Kweichow, and Szechwan:

Province	Sex	Number	Measurements in Centimeters		
			Height	Length	Heart Girth
Hunan (Chang, 1943)	Male	30	115	130	165
	Female	163	110	125	140
Kweichow (Chen, 1943)	Male	40	119	120	167
	Female	157	103	113	138
Szechwan (Huan, 1943)	Male	38	120	130	170
	Female	186	110	120	140

The height was taken at the top of the withers and the length from the shoulder point to the pinbone.

The following measurements were obtained by Hsu (1940) of animals winning prizes in an exhibition in Tzetung, Szechwan:

Sex	Placing	Measurements in Inches		
		Height	Length	Heart Girth
Bull	1	51	53	68
	2	48	51	66
	3	52	57	67
Cow	1	54	54	67
	2	48	53	62
	3	47	49	63
Steer	1	54	60	75
	2	53	56	72
	3	53	61	78

Most of the cattle in Tibet, Sikong, and adjacent portions of the Tibetan plateau are quite small and light-muscled. In general conformation they are more nearly like the Mongolian than the South China draft cattle. However, the conformation and color patterns are similar to hill cattle found on the southern slopes of the Hima-

FIGURE 32.—Tibetan cattle. *Left:* One of the authors (R.G.J.) holds a Tibetan cow. *Right:* A yoke of Tibetan cattle hitched to a native plow.

layas and in northern India and Bhutan, and it is possible that the cattle of the two areas are related.

Data on milk production of native cattle are very scanty. Liang (1943) gives the following average figures on 10 yellow cattle of Szechwan:

Lactation Period (days)			Milk (pounds)			Average Fat
Shortest	Longest	Average	Lowest	Highest	Average	Percentage
27	203	104. 7	30. 5	625. 0	262. 8	5. 9

Management and Feeding

Cattle are used primarily as draft animals on farms and as pack animals in commercial transportation. The water buffalo is the primary source of draft power on farms where rice is grown, although cattle are used to some extent in the rice fields. Yellow cattle are used to a greater extent in the mountainous areas in the southern portion of the country for farm power and are one of the most important sources of such power in the dry-land farming areas farther north. Milk is rarely used in southern China except in the larger cities, but in the northern areas many cows are milked.

The stronger draft oxen can pull a load of 450 to 650 kilos, but the average load is about 200 kilos, and they can cover 15 to 20 kilometers per day. Pack oxen are usually loaded with from 60 to 80 kilos, but the load varies with the size and strength of the animals and the distance to be covered per day.

Shelter for cattle may be provided in the farmer's own living quarters. This is true in the lower Yangtze River Valley, for example, and the quarters are kept quite clean, the droppings being cleaned up at frequent intervals. In some sections it is not uncommon for the farmer and part of his family to sleep just over the cattle, as a protection from theft and possibly as a source of warmth in winter. If separate shelter is provided, as it is in many areas, it is never far from the living quarters and is usually adjacent to them, either at the rear or side of the house, and is walled in by earthen or other fences. In areas where lumber is cheap the animals may stand on slats so that the excreta collect underneath, or they may stand on boards which are covered with leaves or grass to soak up the urine. New layers are put on daily and the whole collection is cleaned out at intervals. In other areas soil is the material most commonly used for bedding. In some areas the barn has a pit in which urine collects. This pit may be at the side of the barn, and in northeastern Hunan it is often built under a wooden floor upon which the cattle stand.

The harness for draft cattle is quite simple. It consists of a bent wooden stick placed in front of the wither to serve as a yoke, with a rope or strap passing from each end under the throat. Traces are also attached to each end which pull the load. There are some variations in the details of the harness in different areas but this is the basic design.

Packsaddles are similar in design to those described for horses and are made in two sections. The base of the saddle is of wood and rests on either side of the backbone, with a pad to protect the skin. This is designed to hold the second portion in place, which is a framework to which the load is attached, so it can be placed on or taken off the animal without moving the base of the saddle. Long sacks filled with grain may be placed across the back without the support of a saddle. This practice is common in the northwest.

Animals are driven almost entirely by voice and whip.

Cows kept for family milk supply in the north or for marketing milk are generally milked twice daily, except in dairies, where they may be milked three times. Many native cows will not let down the milk unless the calf is present and some will not let down the milk unless the calf is first allowed to nurse to start the milk flow. With some, it is necessary to let the calf nurse several times during the milking process, for the cow will let down milk only for a short time after nursing the calf. In such cases the calf is kept tied near at hand and is allowed to nurse just long enough to start the flow of milk, after which it is pulled away. In areas where milk is sold, the calves are kept on starvation diets, being allowed just enough milk to keep them alive to be available to stimulate the milk flow; they are either killed or allowed to die when no longer needed for this purpose. The level of feeding even when they are allowed to grow is usually too low for normal development. In range areas the calves may be tied away from the cows at night, used to stimulate the milk flow in the morning, and after milking allowed to run with the cows all day. They are tied up again in the evening and the cow milked again to get such milk as was not taken by the calf.

In the farming area, the chief roughage during the spring, summer, and autumn is grass and weeds that are cut from along the edges of the fields and from rough areas, such as the many burial grounds where no crops are raised. Since labor is cheaper than fertilizer, this grass is carried to the cattle so the manure can be saved for farm use. During the winter, rice or other straw is fed, along with such green soilage as may be obtainable. During the busy working season such concentrates as corn, horse beans, and bran are fed, and eggs are sometimes given, mixed with other feed. Other concentrates used to some extent for cattle feed in the areas where they are available are soybeans, soybean cake, sesame cake, cottonseed cake, rapeseed cake, barley, kaoliang, millet, soybean milk, and dregs from the making of soybean curd. The amounts of concentrates given in most areas are small.

In the grassland areas the animals live on the range throughout the year. Some supplemental roughage and concentrates may be given to them during seasons of hard work, in the areas where some farming is done, or to pack animals that are used in commercial transportation, but for the most part the cattle in these areas obtain practically all their feed by grazing. Animals that are not working are kept in the lower valleys during the winter and on the higher mountain slopes during the warmer season. Feed during the winter and early spring is inadequate and animals become weak, thus being susceptible to disease and unable to endure undue hardships. If a heavy snow covers the grass for a few days, death losses are heavy.

Dairy cattle are usually barn-fed the year round. Grass is cut during the warm seasons, and straw (wheat or millet in the north and rice in the south) is the chief roughage during the winter. Straw is usually cut into short lengths and mixed with concentrates for feeding. In milder areas, such as around Chungking, it is possible to cut grass throughout the year. Liang (1943) gives some data on the amount of grass that can be cut by the average man in one day. The average for the year was 315.5 catties (catty equals one-half kilo or 1.1 pound). The amount that can be cut is highest in July (411.6 catties) and lowest in February (178 catties).

Liang (1943) also reports the average amount of grass eaten by a cow during a seven-day trial in April. This test included eight cows, and the average consumption per head per day was 102.3 catties. Thus it would require about one man to gather grass for each three cows under conditions in the dairy where the cows were as well fed as existing conditions permitted. The average daily consumption of grass by milking cows throughout the year was 70.6 catties, or 5.53 catties per pound of milk produced during a one-year period. In rainy weather and in the winter when grass was scanty, some rice straw, hay, and carrots were used to supplement hay. Milking cows were fed 9.74 pounds of concentrate mixture per head daily, on the average, or 0.72 pound of concentrate per pound of milk produced during the same period of one year. Dry cows, heifers, and bulls received 4.09 pounds per head daily, on the average.

Grazing of cattle around villages may be handled in three ways. Some individual in each family may be responsible for the cow or cattle belonging to that family. In some villages one man will contract to care for all the animals in the village while they are grazing and is paid so much per head by the owners. In other villages a rotation system is worked out, so that each family or some member of it is responsible on certain days for handling all the cattle in the village while out grazing.

Cattle may be marketed throughout the year, but the major portion are sold at two seasons, spring and autumn. In the spring there is a preponderance of cows and calves on the market, while in the autumn there are more bulls and oxen. In general, more animals are marketed in the autumn than in the spring. During the main marketing seasons a cattle-marketing day occurs every few days. The number of animals for sale in any one market will vary from time to time, and the size of the market will vary in different areas. In Szechwan, there are usually about 100 animals offered for sale on a market day.

On the average, there is one cattle market per hsien (county) in the cattle-producing areas. Sales in these markets are usually arranged by middlemen or commission men who function also as tax collectors. These men collect a certain portion of the selling price, usually about 5 percent, a part of which is tax and the rest is their commission. There is often no set tax in terms of percent. The commission man must pay a certain amount to the local government each year, and any he collects in addition is his income. The commission fee is collected entirely from the seller in some markets, while in other parts it may be collected from both the seller and the buyer. In many of these markets there is every opportunity for the commission man to squeeze. Bargaining is done in the sleeve, and neither buyer nor seller knows exactly what the deal is. Hence the buyer frequently pays more than the seller receives and the middleman gets the difference plus the regular commission. In other areas this dishonesty is avoided by posting the price. The commission man, in addition to setting the price, often insures the animal against death for a short period—in most cases 5 to 7 days.

Some animals are also sold at private treaty, outside the regular livestock markets.

Before the present war, modern dairies existed in the larger coastal and other eastern cities and milk was marketed in a modern manner, with some inspection to guarantee purity of product, from the standpoint of both adulteration and infection. Most dairies in the interior are not well equipped owing to present conditions, and inspection systems have not been developed. Watering of milk is practically an accepted practice, and rigid steps will be necessary to stamp out this extremely bad custom. Deliveries are usually made by men on foot or on bicycles. Bottles are not uniform in size and are frequently under the advertised size, so the customer not only gets a diluted product but less than he pays for. Milk is usually cooled by ice or cold water, depending upon the area. Some dairies pasteurize in order to better preserve the milk but advise customers to boil it before using. Boiling of milk before use is a necessary practice throughout China.

The house-to-house peddling of milk, the cow being milked at each stop, is still a practice in some areas.

A limited amount of butter is made and sold around the larger cities. It is packaged in stoneware jars or wrapped in paper.

In the grassland area there is little marketing of milk, most of it being used for drinking or for the production of butter and casein. Butter is made in simple wooden churns, or by putting the milk inside a cattle or yak skin and rocking it back and forth until the churning is complete. The butter is then stored in wooden containers for winter use or may be packed in sheep or goat stomachs for marketing. It is very rancid before reaching a market. Much of the butter in the Tibetan areas or areas where lamaseries are found goes to these lamaseries, where it serves as fuel in the candle pots. Lamas also use butter for carving many fancy images that are used for one day at a ceremony in the early spring. Much butter is also used for mixing with tea for drinking in Tibetan areas. Casein is made from the milk left after churning, by boiling to coagulate the protein and then putting this material out in the sun to dry. Most of this is used for food during the winter and the whey is fed to the calves. Butter is also used by Tibetans as a skin lotion to protect against the wind and to treat sores and burns, to soften animals skins, as a water-repelling agent on clothing made from textiles, and as a hair dressing.

Part of the leg bones of cattle are used for toothbrush handles. Six bones are suitable for this purpose, and 4 handles can be made from each, or a total of 24 from each animal. The other bones are used for other types of articles made from bone or are pulverized into bone-meal.

Hair from cattle is woven into ropes and is also used for making blankets.

Skins are used for leather and may be tanned with tannin from natural sources or with certain chemicals. Smoke-tanning is also done in some areas, by smoking the individual skins over fire boxes which are stoked from the top with straw to produce a maximum amount of smoke. Cattle skins from adult animals average about four kilos in weight when dried.

Breeding

Little is known of the age of puberty. This is conditioned to some extent by the plane of nutrition, and it is generally believed that the cattle of northern China reach puberty a little later than those of the southern portion, and that males reach puberty a little later than females. Bulls are said to be ready for service at 2½ years. Some females are reported to show estrus at 10 to 12 months of age but are seldom bred before 14 to 20 months.

Signs of estrus are not usually marked, and it is quite difficult for the farmer who has only one cow to know when she is in estrus. Definite data on the length of estrus and the estrous cycle are not available, but the duration of estrus is reported to be about 1 to 2 days and the estrous cycle from 18 to 21 days, and sometimes shorter or longer.

The length of pregnancy is reported by Chen (1943) to vary from 268 to 309 days in Kweichow, while Huan (1943) gives a range of 270 to 285 days under southeastern Szechwan conditions. Liang (1943) gives the average of 26 gestations in native cows as 278.3 days with a standard error of 1.4 days.

The breeding season is quite variable. Most animals are either bred in the spring or autumn, with summer ranking next, and the smallest number are bred in the winter. In the grasslands and in some farm areas the males run with the females and breeding is done whenever a female is in estrus. In many farm areas hand-mating is practiced. In such areas one member of a village usually keeps a bull to which cows are brought, and a charge is made for the service.

Most bulls that are to be worked on farms are castrated. This operation is sometimes performed at 6 months but may be delayed to as late as 4 years. Castration is done, without any antiseptic precautions, by the use of a sharp knife. An incision is made either lengthwise or crosswise to the long axis of the scrotum and the testes squeezed out. The cord is cut with knife or scissors.

In some areas eggs are fed to bulls when giving service. Two eggs per service may be given. In Hunan, it is reported that some farmers feed bulls small fish and shrimp to supply protein. There is also a practice in some areas of giving the cow wine after each service, with the belief that she is more apt to settle if this is done.

There is evidence in other countries (Phillips, *et al.*, 1943) that the level of fertility may decrease during the summer months, owing to the high environmental temperatures, and there is evidence that a similar condition exists, at least in southern China. Liang's data (1943) from Chungking are as follows:

Items Compared	March-May *Spring*	June-Aug. *Summer*	Sept.-Nov. *Autumn*	Dec.-Feb. *Winter*
Number of services	192	152	241	210
Number of conceptions	98	64	145	123
Percentage of fertile matings	51	42	60	59

In many areas in China it is reported that cows frequently come in estrus during pregnancy. Liang observed this phenomenon three times in 26 pregnancies in native cows, or in approximately 12 percent of the cases.

Practically all the cattle imported into China have been of dairy type. According to the 1936 Chinese Yearbook, the numbers of dairy cattle in the country were as follows: Holstein, 2,607; Ayrshire, 171; Guernsey, 138; Jersey, 102. Other breeds have been imported in very limited numbers. There is a report that some Brown Swiss bulls were recently brought to Sinkiang from Russia.

Before the outbreak of the present war the dairy industry had made some progress in Nanking, Tsingtao, Tientsin, Peking, Canton, and other cities, and especially in Shanghai. At the outbreak of the war the cattle belonging to Central University and to Revolutionary Heroes Descendant School were moved to Chungking and Chengtu, respectively. The latter herd is now maintained by Madame Chiang Kai-shek, under the supervision of Frank Dickinson, West China Union University, Chengtu. Dairies in other cities were destroyed or confiscated by the Japanese. Considerable progress has been made in Chungking and Chengtu and in some other areas during the war.

Purebred dairy animals have been used in commercial dairies and by a few individuals, and purebred bulls have been used for grading up native cows in some communities. Few data have been recorded to show the results of this grading up, or the performance of the purebred animals under Chinese conditions as compared with the conditions in countries from which they were imported. Liang (1943) gives some data on this problem and these are summarized below. Body size as indicated by various measurements was determined on a small number of native cows and on grades in the herd, and some of the averages are as follows:

Item	Yellow Cow	½ Holstein	¾ Holstein	Holstein
Number of animals	7	41	3	4
Height at withers, in cm	110.6	121.0	123.3	126.5
Length, poll to pinbone	166.3	191.3	189.7	208.3
Heart girth	151.4	173.7	175.3	182.3

In general appearance the first-generation grades possess much of the beefiness of the native cows, but later generations are more refined and similar to the Holstein in appearance. Birth weights are not available on native cows, but there has been a progressive increase in size as the amount of Holstein blood in the grades increased, as indicated by the following figures for birth weight:

Breeding of Calf	Males		Females	
	Number	Weight	Number	Weight
		Lbs.		*Lbs.*
½ Holstein	11	54. 8	8	52. 7
¾ Holstein	19	71. 6	44	67. 4
⅞ Holstein	8	74. 4	9	67. 3
Holstein	8	88. 2	5	73. 1

The color of Holstein grades is of some interest and the results in the Chungking Dairy may be summarized briefly as follows. Of 50 first-generation cows, none were all black but 49 showed more black than white, and 1 was approximately half black and half white. Of 56 second-generation animals (¾ Holstein), 2 were black, 32 had more black than white, 7 were about half and half, and 15 had more white than black. Nine third-generation animals (⅞ Holstein) were divided as follows: 3 with more black than white; 3, half and half; and 3 with more white than black. Three brown animals have occurred in the first generation since this tabulation was made. These results are not necessarily typical of what would be expected with other native cattle or with other foundation Holstein stock.

Dickinson (1943) has one three-fourths Holstein female in his herd that is reddish yellow and white, and animals of similar color have occurred in the grades in the Central University herd. These cases may be due to the presence of a recessive red-color gene in the bulls used for grading up.

615999—45——5

The production of Holsteins, natives and grades, as reported by Liang (1943) is summarized below:

Breeding	Number	Lactations	Lactation Period			Milk Yield (Mature Basis)	Fat (Percentage)
			Long	Short	Average		
Yellow cow	10	13	203	27	104. 7	[a] 262. 8	5. 92
½ Holstein	54	166	530	180	329. 8	5, 152. 8	4. 74
¾ Holstein	20	27	575	259	386. 1	6, 567. 3	3. 90
Holstein	9	17	639	276	418. 2	9, 196. 8	3. 63

[a] Range was 30.5 to 625.0 pounds.

The lactation periods are given in days and the milk yields in pounds. The fat-percentage figures are for slightly different groups of cows, including 10, 52, 29, and 6 cows; and 10, 434, 230, and 53 samples, respectively. More data are needed as a basis for determining production of native cows and of higher-grade cows.

Hsu (1943) collected some data on native cattle and on Holstein × yellow-cattle crosses in Kwangsi. Animals were tested at 10-day intervals for a period of 2 months during November and December. These animals calved during the previous months of August and September, with the exception of one Kwangsi yellow cow that calved the previous March. The grade Holstein cows were better fed than the native cows. Three yellow cows from Kwangsi averaged 9.1 percent fat while three from Kiangsi averaged 7.9 percent. Two Holstein grades averaged 5.8 percent. The average daily milk production for four testing days (animals were milked twice daily) for the individual animals was 515, 530, and 675 cc. for the Kwangsi yellow cows; 444, 697, and 1,085 cc. for the Kiangsi native cows; and 2,651 and 4,835 cc. for the Holstein grades.

According to one report (Wang, 1942) Holsteins were being used for grading up in Manchuria. No results are given.

Some data that are of interest on level of fertility have been accumulated by Liang (1943). These are summarized below:

Breeding of Cows	Number Services	Number Pregnancies	Average Number Services per Conception
Yellow cows	36	26	1. 38
½ Holstein	255	165	1. 55
¾ Holstein	58	28	2. 07
Holstein	43	20	2. 15

The length of gestation was similar in all groups of cows, the averages for the four respective groups listed in the above table being 278.3, 278.1, 275.4, and 277.1 for 26, 154, 28, and 16 pregnancies.

The results summarized above indicate that the Holsteins and their grades are performing satisfactorily under conditions in Chungking, from the standpoint of milk production. Fat percentage is reduced as compared with the natives, but this must be anticipated if the level of milk production is to be increased with a breed adapted to market milk production like the Holsteins.

Careful observations need to be made under other conditions to determine if the purebreds and grades are standing up well under the local environments. Observations on fertility indicate that the yellow cattle may be superior in fertility, and this point needs careful study as grading-up programs progress.

FIGURE 33.—Results of grading up yellow cattle of Szechwan with Holstein and Jersey bulls. *Top: left,* A native cow; *center,* A ½ Holstein–½ native cow; *right,* A ¾ Holstein–¼ native cow. Maximum daily productions of the three animals were 7, 20, and 33 pounds of milk, respectively. Dr. Frank Dickinson of West China Union University, Chengtu, Szechwan, who has done much pioneer work in the introduction of improved plants and animals, holds the animal at the right. *Bottom:* A first cross Jersey × native heifer produced at the West China Union University. The Jersey is not nearly as popular as the Holstein in China, for in the words of one worker there: "It takes too much water to adulterate Jersey milk."

Water Buffalo

The water buffalo, as its name implies, likes water and is able to walk with comparative ease in mud. It is also a quiet-tempered animal. These characteristics make it especially well adapted for use in the rice field, and its distribution in China is limited almost entirely to the rice-producing area. The main use of the water buffalo is for draft power. It is used to a limited extent for milk production in only a few localities.

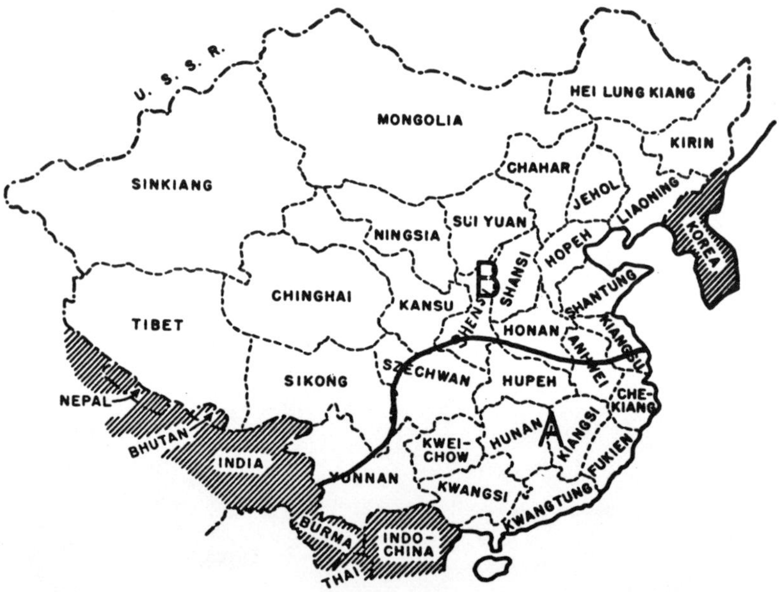

FIGURE 34.—Distribution of water buffalo in China. *A*, Area in which water buffaloes are used extensively. *B*, Area in which no buffaloes are used, except for a few in localities adjacent to area *A*.

Types

There is considerable variation in conformation and in color but there are no types that are recognized as being distinct. The most common color is usually characterized as black, but it is more nearly a dark slate-gray. There are some tan or "reed-flowered" animals and an occasional white one. Farmers prefer the black ones since they are believed to stand the cold better and be easier to feed in the winter. White ones are considered to be unlucky in most regions and are, therefore, found in small numbers in certain limited areas only.

FIGURE 35.—The water buffalo is the most important source of power on farms in the rice areas of southern China, especially because of its ability to work in the mud and water of flooded rice fields. *Top:* A typical water buffalo of China, photographed in Kweichow. *Middle:* White water buffalo, but not an albino. White animals are infrequent, and are generally not favored by farmers. *Bottom:* A water buffalo pulls a harrow through a rice field in Szechwan.

The water buffalo is of rather heavy, beefy build, on the average. The body is heavy, the chest is deep and fairly wide, the legs short, and the middle large. As compared with cattle, it tends to be somewhat pot-bellied. The bones of the leg are large and fairly strong, although cow-hocked or sickle-hocked hind legs are rather frequent. The chief differences in conformation between it and cattle are the more sloping rump in the water buffalo, along with a shorter tail-head, and the angle at which the head is attached to the neck. This angle is such that the face is nearer the horizontal plane, while in cattle it is nearer the vertical plane.

The horns differ in shape and direction from those of the cattle. They are angular instead of round at the base and extend upward and backward over the neck. There is considerable variation in the direction and degree of bending. In some animals the horns almost form a circle, so that the ends are only about five inches apart, while at the other extreme, the horns extend more to the side and the tips may be two feet apart. Farmers are said to prefer the coiled type in cows and the straighter type in bulls.

Hoofs are of two different shapes. One is called the scissor type, in which the toes are long and inconvenient for walking. There is an undue amount of wear on the bottom of the foot, resulting in tenderness and limping. The other type is called the bowl type. The foot is square set, with the pastern at about a 45-degree angle. The hoof wears more evenly, and this type is favored by farmers. Three types of hoof quality are recognized. The most desired is that in which the horn of the hoof is black, since this type has been observed to be tough and to wear well. A second type in which the horn is yellow or waxy in color is not so hard and does not wear as well. The third type is composed of a mixture of black and yellow horn and is said to be intermediate to the first two types in hardness.

Some data are available on the sizes of animals in certain provinces, as indicated by body measurements. Data from areas in which are located three of the Ministry of Agriculture and Forestry's cattle stations are summarized below:

Province	Sex	Number	Height (*cm.*)	Length (*cm.*)	Heart Girth (*cm.*)
Hunan (Chang, 1943)	Male	47	130	135	180
	Female	69	121	129	170
Kweichow (Chen, 1943)	Male	45	131	140	193
	Female	86	120	127	174
Szechwan (Huan, 1943)	Male	51	133	143	194
	Female	81	123	130	175

The length was measured from the shoulder to the end of the pin-bone.

Hsu (1940) gives the following measurements on the prize-winning animals at an exhibition held in Tzetung, Szechwan:

Sex	Placing	Height (inches)	Length (inches)	Heart Girth (inches)
Bull	1	51	55	79
	2	52	56	75
	3	53	57	79
Cow	1	53	57	78
	2	52	55	82
	3	51	57	75
Steer	1	57	62	86
	2	56	55	83
	3	53	62	82

Levine (1920) states that water buffalo in the Canton area range from 46 to 56 inches (116.8 to 142.2 cm.) in height, and 800 to 1,400 pounds (364 to 636 kilos) in weight.

Few data are available on the milk and fat production of water buffalo. Hsu (1943) gives data on six animals in Kwangsi. These data were collected during November and December on animals that had calved during August and September of the same year. Samples were obtained at intervals of 10 days for 2 months for tests of percent butter fat, and these averaged 11.0 percent. The average daily milk yield of 4 animals (animals were milked twice daily) at 4 successive testing periods was 605, 828, 964, and 2,317 cc., respectively.

Levine (1920) states that a water-buffalo dairy was started in Canton in 1905 and that observations indicate a promising future if a program of improvement in milk production is undertaken. Verbal reports indicate that other buffalo dairies were in operation around south-eastern coastal cities prior to the war but no data are available on production.

Management and Feeding

The primary use of water buffalo is as a source of power on farms, and especially on farms in the rice-producing areas. In Szechwan a considerable number are used in the pumping of brine from the salt wells. Some are also used as a source of power for the grinding of sugarcane. They are used to a very limited extent for milk production as a by-product of work. Surplus animals and animals that are no longer suitable for work are sold for meat.

Data from various sources indicate that the buffalo on the average farm is not used more than about 40 to 50 days per year for farm work.

Methods of management are similar to those described for cattle. There are certain differences owing to the buffalo's liking for water. Where small ponds are available, several animals may be tied around it when not in use or during the night so they can lie in the water and mud. The owner then takes them to a large pond or stream where they can clean off in clear water. Or the buffalo is frequently taken to a stream at the end of a day's work to allow it to enter the water and cool off.

Feeds and feeding practices are similar to those described for cattle.

By-products, such as bone and skins, are used in the same manner as those of cattle. The water-buffalo hide is larger, thicker, and heavier than that of cattle, weighing about 13 catties (6.5 kilos) when dried, for mature animals, as compared with 8 catties (4 kilos) for cattle.

Breeding

Breeding practices are similar to those outlined for cattle.

Little information is available on reproductive phenomena in the water buffalo. Observations indicate that females may first show estrus sometime between 12 and 18 months of age, but breeding is usually delayed until about 2 years of age. Observations also indicate that estrus may last for 2 to 3 days, and that the estrous cycle varies from 18 to 28 days. Actual data are needed on these points. The male is considered ready for service at 3 years but no data are available on the actual age of puberty. Data on pregnancy indicate that it is somewhat longer than in cattle, varying from 300 to 348 days. More accurate information is needed.

Testes of the male are relatively smaller than those in cattle. Slaughter data obtained by Hsu (1943) show that the average size of the testes, including epididymis and a little other tissue, was 0.069 percent of body weight in water buffalo and 0.106 percent in cattle.

Yak and "Pien Niu"

The yak is indigenous to the mountainous regions of central Asia, and wild herds are still found in some regions. It is well adapted to grazing on steep mountain slopes and for travel in mountainous regions. The *pien niu*, or hybrid between yellow cattle and yak, is found in practically all the regions where the yak is used. The yak is related to the American bison and is similar to it in many respects. It grunts with a sound more like that of a hog than like the lowing of cattle of western countries.

Types

There are no distinct types of yak, but there are wide variations in size and color and also some variation in conformation.

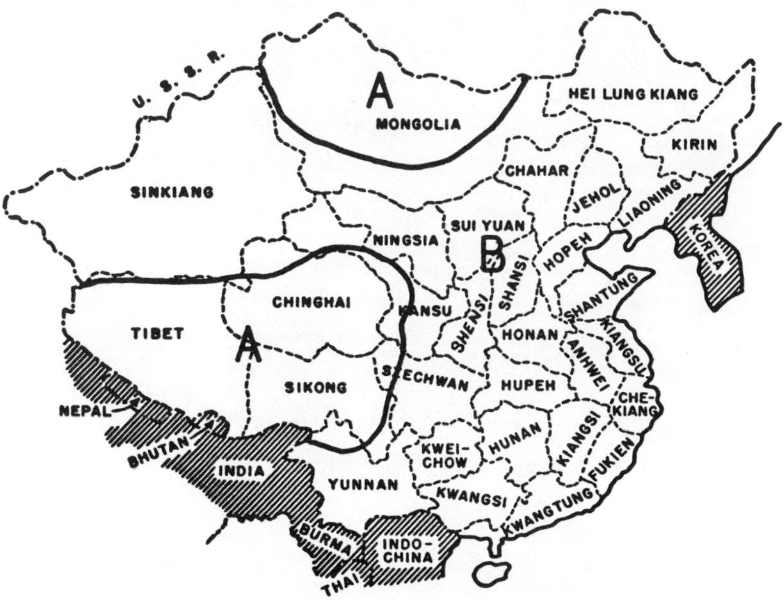

FIGURE 36.—Distribution of yak in China. ***A*, Approximate areas in which yak are commonly found. There is not sufficient information at hand to define these areas very precisely. *B*, Area in which yak are of little or no importance. The distribution of *pien niu* (cattle-yak hybrids) is similar to that of the yak.**

Figure 37.—Yak are used extensively in Tibet and Outer Mongolia for packing and for other work and for milk. *Top:* Bull belonging to a lamasery in Lihua, Sikong. *Middle:* Yak cow and calf. *Bottom:* A good specimen of a young yak cow, still carrying her winter coat.

Female yaks in Sikong are estimated to average about 109 cm. in height and 225 kilos in weight. Kislovsky (1938) gives the average height of female yaks in Outer Mongolia as 109 cm. Males are larger on the average and some become quite massive. The *pien niu* is described by Kislovsky as being larger in all its measurements than either of the parent types and weighing about 18 percent more than cattle. In Sikong it is estimated that *pien niu* are approximately 20 percent larger than yaks.

The yak is not specialized in conformation for either beef or milk production. The legs are rather short and the body relatively heavy. The head is narrow and long, with the forehead relatively wide, and the face usually dished. The front portion of the back is higher than the neck and slopes off toward the rear. The rump is usually sloping, and the tail is short and covered with long hair resembling the tail of a horse more than that of cattle. The chest is deep in relation to height, but the rear quarters tend to be light. The horns are slender and long and curve outward and upward. Polled animals are also frequently seen. In Outer Mongolia polled animals are preferred, according to Kislovsky, and no horned bulls were observed, all the horned males having been castrated. The hoofs are usually big and strong and well adapted to mountain-climbing or to walking on marshy grasslands. The hair of the yak is longer than that of cattle. Fringes around the poll, at the withers and shoulders, along the flanks and on the thighs grow quite long, giving the animal a very hairy appearance. From this characteristic it receives its Chinese name of *mao niu* or "hairy cattle".

There is considerable variation in the color of yaks. Black, dark brown, and black with white spots are all frequently seen. Gray and white animals are relatively less frequent. The wild type of pigmentation or agouti is occasionally seen. Kislovsky (1938) states that about 20 percent of the animals in Outer Mongolia are of solid color, and the remainder show varying degrees of white markings, ending in practically pure white with a few colored specks. A type of roaning which gives the animal a blue tint was observed in 21.5 percent of the cases. Two types of white markings were described. In one, the white is limited to the head, tail, feet, and inguinal region. In the other, the "color-sided" type of piebaldness is found.

The *pien niu* is intermediate to the yak and cattle in conformation, and, as has already been mentioned, is larger and stronger, exhibiting the phenomenon of hybrid vigor. The hair is also intermediate in length. Colors are similar to the yak, but some brown pigmentation is seen, particularly the lighter browns. These shades are seen in animals that are frequently classed as yaks but which carry some yellow-cattle blood.

Little data are available on the milk-producing ability of the yak. Records being collected at the Pa Mei Livestock Farm in Sikong indicate that under native conditions of feeding and management the yak produces 2.5 to 4 pounds of milk daily for a period of four or five months. Some of the better ones may give a little more and for a longer period. The milk is of a rich golden color and the butter also has this color. Kislovsky (1938) states that the milk yield of the yak is low, but that the fat content is high—6.34 percent, which is apparently an average figure.

FIGURE 38.—*Upper left:* A yak serves as a mount for a nomad caravaner in Tibet. *Upper right:* Yak plow through the snow on a plain north of Tsange La, Tibet. *Lower left:* A yak caravan in Sikong. *Lower right:* A yak pulls a native cart in Tibet. Note the pack-saddle base adapted for use as harness.

The *pien niu* produces more milk than the yak and is generally believed to produce more than yellow cattle. During the early part of lactation the daily production is about 6 pounds, and some individuals may give 10 to 12 pounds. Huang (1943) reports that an exceptional animal gave 20 pounds daily. The average daily production in the herd in which this animal was maintained was 5 pounds for the lactating animals, and they were milked for about 7 months on the average. Most of the 100 animals in this herd were purchased in Chinghai. Tests of fat percentages made on four animals in Sikong by Lu (1943) showed an average of 5.6 percent. These animals had freshened in the spring and the data were obtained on July 1.

Management and Feeding

The yak and the *pien niu* are used for packing in commercial transportation and also for the carrying of wood and supplies by private individuals. In addition to their use as beasts of burden, they produce milk and thus provide one of the staple sources of food in the high plateau regions. They are also used for meat, and the hides are used for leather. Hair is used for making rope and smaller thread for weaving tents and for other purposes.

For the most part the animals are grazed in the vicinity of the villages, or of the nomad tents during the summer months. No supplemental feed is given during the season when grass is obtainable, and no hay is put aside for winter use. In periods of heavy snowfall, especially during the later winter and early spring, losses are frequently heavy, since resistance is low and animals cannot survive many days without feed. In such periods, choice animals are sometimes fed dried casein or other products to keep them alive. In many areas, the ranges are overstocked and there is urgent need for better range management. The more enterprising owners allow their animals to graze both day and night, and in areas where it is unsafe to allow the animals to graze alone at night, a guard spends the night with them. Some owners allow grazing only during the day and tie up the animals at night.

New-born calves are housed in the yak-hair tents during the night, for safety and also to reserve all the morning milk for human use. The more considerate milkmaids may leave a little for the calves. The calves then follow their dams during the day unless the weather is too stormy. At night the milk not taken by the calf is obtained for human use. In some areas animals are milked only once daily, usually in the morning. Some yak cows are not milked when producing a *pien niu* calf, in order to let it develop more fully.

The practice of allowing the calf to suck just enough to stimulate the letting-down of milk is almost universal among the nomads. It appears to be essential with yaks, but many *pien niu* will apparently let down their milk without this stimulation. With some females, it is necessary to allow the calf to suckle several times during the milking process to obtain all the milk. Skins of dead calves are sometimes stuffed with straw and used to help stimulate the milk flow of the dam.

The Tibetans usually milk with the hands wet. Dry-hand milking apparently has not been practiced in this area. The front legs, and also the rear legs, are usually tied together while yaks and *pien niu* are being milked.

Concentrates are seldom fed to calves. Some whey left from the making of casein or "milk residue" is occasionally fed. Calves learn to eat grass when quite young. During the summer months when grass is growing the calves make fair growth, but during other portions of the year, and especially during the late winter and spring months, there is not enough feed to permit normal growth. These periods of near-starvation stunt the young stock and weaken mature animals. Losses are heavy. It is especially difficult for the calves and yearlings to "drag through" the winter.

Yak and *pien niu* packsaddles are of the same design as those used for horses and cattle. Loads are removed at the end of the day or trip, but the saddle is frequently left on for days at a time, apparently owing to a desire to avoid the difficulties of getting animals saddled each morning, but sometimes to protect the animals from storms. Saddle sores occur frequently.

Most animals and animal products are used by the people that produce them. Some animals are sold for use in pack caravans but there is no organized marketing procedure. In some areas in the grasslands, markets are held at certain seasons in connection with religious festivals. Milk is used for drinking and for production of butter and milk residue or casein. Some of these products go to the lamaseries and some is stored for winter use. Relatively little is sold. Butter for marketing may be packed in sheep or goat stomachs. Hair is collected as it falls out or when the animals are shorn during the summer. Yields are said to be about 1 to 1.5 pounds per animal but no actual data are available. This is used for making tents, felt, ropes, and other things. The coarse hair is used for twine from which tents, grain bags, and similar articles are woven, and the finer under coat is used for felt. Tail hairs are collected and used for the making of dusters.

Some yak cows and a considerable number of *pien niu* cows are kept around the cities in and near the grassland areas for the production

FIGURE 39.—*Left:* A yak skin is used for a churn. It is filled with milk as it accumulates and rolled about until the butter is formed. Tibetans prefer rancid butter. (Photographed by T. Y. Hsu, Chinese Ministry of Agriculture). *Right:* Coracle or skin boat made from yak skins.

Figure 40.—Yak hair is spun into thread and woven into strips of cloth, which are sewed together to make tents. *Top:* One of the two largest tents in Chinghai Province. These belong to Governor Ma Pu Fang and are located near Sancheocheng, (Triangle City). They are equipped with permanent stoves and beds. *Bottom:* A nomad family's tent. Note the casein drying in the sun at the lower right.

of market milk. Most of these are small individual enterprises, but a few have been organized on a commercial scale. Milk is also purchased in some areas for the production of casein for glue, largely a wartime development for airplane-lamination work.

Some hides are sold for leather, but the tanning and processing of leather is generally poorly done. In Sikong, a considerable portion of the hides are used in the raw state to wrap tea, butter, and salt for transport. The material is wrapped with the wet hide which deteriorates in transit so that usually only one service is obtained from each hide, which is a very wasteful procedure.

Breeding

No data are available on the age of puberty in yaks and *pien niu.* The yak calf is very small at birth and is estimated to weigh only 8 to 15 kilos. Growth is slow, owing perhaps to both heredity and environmental conditions, and mortality is high. The *pien niu* calf is somewhat larger than the yak calf, on the average, and growth is believed to be somewhat more rapid. Much of the observed slow growth and high mortality in calves may be owing to a common practice of starving calves in order to have milk for the family.

No data have been published on estrus or the estrous cycle or on the duration of pregnancy. Huang (1943) reports that the estrous cycle and the period of pregnancy in the yak and *pien niu* are similar to those in cattle, or about three weeks and nine months, respectively.

Breeding efficiency is apparently low, although no accurate data are available. In Sikong, for example, it is estimated that not more than 50 percent of the yak females of breeding age produce calves in any one year.

There is no organized system of breeding. Bulls usually run with the females of all ages throughout the year, hence neither the time of calving nor the type of mating can be controlled. In addition to the bulls that run with the owners' herds, there are many herds of "liberated bulls" that roam about at will. These are bulls that have been donated to Buddha and nature, and it is not uncommon to see a herd of 15 or 20 of them roaming about. In the wild state, the time of breeding is reported to be during July and August when the grasses are best, but no accurate data are available. The yak is said to come in estrus about a month earlier than the *pien niu*.

Little selection of breeding stock is practiced. The females in the breeding herd are those that are able to survive the rigorous environment. Selection of a male for breeding may be merely a matter of chance or be based on the owner's preference for some color or other trait. The woman of the family frequently makes the selection of a male to be saved for breeding, and there is one herd near Tai Ning, Sikong, in which most of the animals have white faces, owing to the preference of the wife of the owner for white faces and the continuous selection of white-faced animals for breeding.

Owing to the lack of a systematic breeding program, the breeding of yaks and *pien niu* is in a chaotic condition. *Pien niu* are usually produced by mating yellow bulls to yak females, but offspring from the reciprocal cross are also called *pien niu*. Hybrids from the two methods of crossing are said to differ in certain characteristics. It is claimed, for example, that hybrids from yellow-cow dams have smaller tails and louder voices than those from yak dams. The latter type are said to have tails more like those of the yak and to have longer hair, more quiet disposition, and greater prolificacy. Whether these beliefs have any basis in facts remains to be determined. It seems probable that much of it is folklore that may not stand up under critical tests.

Terminology for describing the various crosses is confused. In some areas the progeny of a cross between the *pien niu* female and either the yak or yellow bull is referred to as a *galiba*, while the Tibetans refer to these as *ah gohr*. In other areas only the progeny resulting from mating yak bulls to *pien niu* females are referred to as *galiba*. This latter cross is the one most frequently made. After a few crosses back to either yak or yellow bulls (three in some areas) the offspring are referred to as yak or yellow cattle, according to the type with which the hybrids were backcrossed.

The *pien niu* males are reported to be sterile. This belief is so widespread in the grasslands that it is apparently true, even though no clear-cut, experimental evidence is available on this point. Kislovsky (1938) makes the same claim concerning *pien niu* in Outer Mongolia.

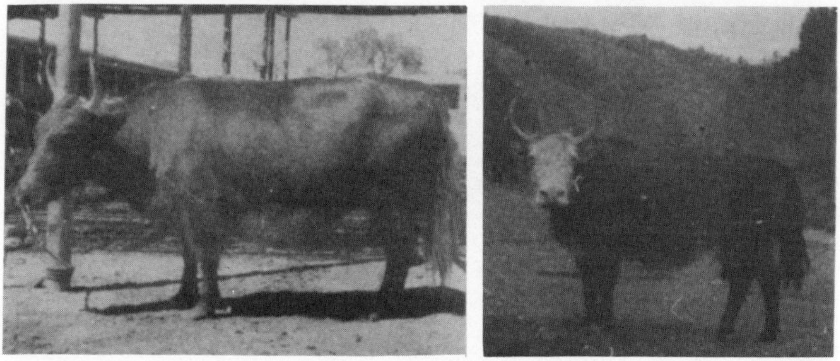

FIGURE 41.—*Pien niu* (hybrids between cattle and yak) are usually produced by mating "yellow" bulls to yak cows. *Left: Pien niu* cow in a dairy in Lanchow, Kansu. *Right:* A *pien niu* in Sikong. White faces such as this are seen occasionally.

The male offspring from backcrosses to either yaks or yellow bulls is also said to be sterile. No problem of sterility or lowered fertility is reported with the hybrid females or with females resulting from backcrosses to either parent race. Zawadowsky (1931) gives some information on the conditions of the testes in hybrids between zebu and yak. These are of interest here since cattle and zebus are interfertile and there has been no report of sterility in the male hybrids. His report indicates that hybrids between zebu and yak have normal sexual desires but that the testes produce no spermatozoa. Only Sertoli cells and spermatogonia are present.

The yak male has small testes and a smaller, tighter scrotum than the male of the yellow cattle. Male yaks, except those retained for breeding, and all *pien niu* and other crossbred males are castrated.

The *pien niu* is generally recognized as being superior to yaks and cattle in size, working ability, and milk production, although few clear-cut data are available to support these popular views. Backcrosses to either yaks or yellow cattle are generally considered to be unproductive and inferior to either of the parent types. Kislovsky (1938) states that there is some heterosis in the first backcross, but that in Outer Mongolia later generations are considered to be unproductive.

The one point that is perfectly clear concerning the breeding of the yak and *pien niu* is that there is urgent need for careful experimental work to determine the type of cattle bull best adapted to producing *pien niu* and the best method of utilizing the hybrid females in a continued breeding program.

Sheep

Types

The sheep of China may be divided into three main types, Mongolian, Tibetan, and Kazakh. There is considerable variation within each type and several subtypes exist, particularly among the Mongolian sheep. There is some overlapping and mixing of types along the borders of the areas occupied by the respective types, especially in southwestern Kansu and eastern Chinghai, where all three types overlap.

Mongolian sheep are a fat-tailed type with coarse wool. The tail fat is deposited mainly toward the base end, while the extreme end of the tail is small and is usually curved or crooked. The predominating

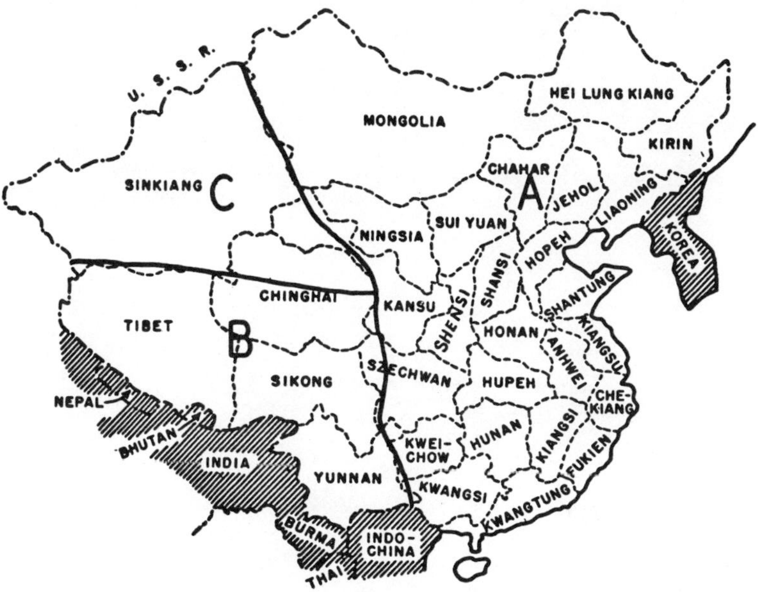

Figure 42.—Approximate distribution of types of sheep in China. *A,* Area in which Mongolian sheep are found. Most sheep are in the northern portion of this area. There are relatively few sheep in South China. A type with especially large tails is found in Shensi and Shansi and other subtypes are recognized in other localities. *B,* Area in which Tibetan sheep are found. *C,* Area in which Kazakh sheep are found. There is considerable overlapping along the borders of these areas, particularly in southern Kansu and eastern Chinghai.

color pattern is white with a colored head, either black or brown. Piebaldness of various types and self-colored animals are also seen. Kislovsky (1938) gives the average live weight as about 48 kilos for adult ewes and 56 kilos for rams. The height at withers is reported to vary between 62 and 68 cm. Wool yield is reported to vary from one to three pounds per year. Males are usually horned, and females may be polled or have small horns.

There are many variations in the Mongolian sheep. Kislovsky (1938) states that there are eight subvarieties in Outer Mongolia, apparently having some admixture of blood from fat-rumped sheep.

FIGURE 43.—Large flocks of Mongolian sheep graze on the open grasslands, while smaller flocks are kept in the farming areas. *Top:* This flock has been brought inside a compound wall for protection at night. *Bottom:* A flock grazing in a stubble field. The "basket" on the shepherd's back is used for watering the sheep. It is filled from a well and used as a watering trough.

FIGURE 44.—Mongolian sheep of Shansi, Honan, and Shantung. *Upper left:* A good male specimen from the Taiku area. Sheep of this general type are found throughout Shansi. *Upper right:* Left to right, these ewes are, respectively, from the Shouyang, Taiku, and Luan areas in Shansi. The Shouyang sheep have a tail that is heavy at the base with a long, fairly straight tip. The Luan sheep have a very heavy tail with a small, up-turned end. *Lower left:* Ram of the Shouyang area. *Lower right:* Hanyang sheep of Honan and Shantung.

All the Mongolian sheep have heavy, fat tails, but two types found in Shensi and Shansi, respectively, have especially large tails. These are the Tung or Tungchow sheep of Shensi and the Hanyang sheep of Shansi. They are very similar in appearance and probably originated from the same basic stock. These animals are of fairly good mutton conformation and are characterized by a very large, pendulous, fat tail. It is somewhat pear-shaped, having two lobes. It hangs down on the hocks of the animal and is a hindrance to grazing and in breeding. Tails often weigh as much as 8 to 10 kilos and sometimes more. Fleece yields apparently average around 1 kilo, but yields as high as 2 kilos are reported. The wool is somewhat finer and more crimpy than many other Chinese wools. Both rams and ewes are hornless.

Several distinct types are recognized in other portions of China. These include the Shantung, Tan, and Woozie (Wusih). These are all similar to the other Mongolian sheep and vary only in certain respects. The Shantung sheep have somewhat larger bodies and the fleece is said to be somewhat finer. They are found in Shantung and the adjacent provinces of Honan, Hopeh, and Shansi. The Tan sheep has somewhat finer wool that is said to be of more uniform fineness.

Figure 45.—Mongolian sheep. ***Upper left and right:*** **Two rams from the northeastern portion of Kansu Province.** ***Lower left:*** **Tung sheep of Shensi Province. It is claimed that "pure" sheep of this strain have a two-lobed tail with no tip turning up.** ***Lower right:*** **Tail of a Tung ram.**

It is found in Ningsia. The pelts are valued for making linings for coats. Woozie sheep are found on the southern side of Lake Tai, in Kiangsu, and Chekiang. The body is larger than that of other Mongolian sheep and the fleece is white and of somewhat better quality than that of Mongolian sheep but not so fine as that from Shantung. The fleeces from these sheep yield about 70 percent clean wool. Before the war the best quality wool was obtained from Shantung according to Hsu (1943). The spring clip was about four inches and the fall clip about three inches in length. Fleeces weighed from two to three catties.

Two other local types of Mongolian sheep, called Taiku and Shouyang, are recognized in Shansi and it is probable that local subtypes exist in other provinces.

FIGURE 46.—Tibetan sheep usually run in flocks on the open grasslands. In the border between farming areas and grasslands small farm flocks are kept. *Top:* Flock on grasslands in Sikong. *Bottom:* Farm flock in a river valley in eastern Chinghai. Note the small tails that are typical of Tibetan sheep.

Tibetan sheep, as the name implies, are found primarily on the Tibetan plateau and adjacent regions. Color markings are similar to the Mongolian sheep, and the head and sometimes other portions of the body may be either black or brown. Both males and females are usually horned and there is considerable variation in size and shape of horns. The wool is similar to that produced by Mongolian sheep but is somewhat coarser on the average and the fleeces are somewhat lighter. The most distinct difference between Mongolian and Tibetan sheep is in the form of the tail. The Mongolian has a fat tail, while the tail of the Tibetan is naturally small and not fat. It is about 10 to 12 cm. long and is straight and conical in shape. There is considerable variation in size and form, but there is not sufficient knowledge of these sheep available to justify classification into subtypes at the present time.

Kazakh sheep are apparently native to Kazakhstan and adjacent areas, from whence they have spread into Sinkiang and adjoining regions of China. These sheep are taller and larger-framed than most of the Mongolian and Tibetan sheep. Color patterns are similar, and the head and other parts of the body may be either black or brown. Males are usually horned and females usually hornless. The

FIGURE 47.—Individual examples of Tibetan sheep. *Upper left:* Tibetan ram of Sikong in full fleece. *Upper right:* Tibetan ewe in Lhasa, Tibet. *Lower left and right:* Tibetan wethers of eastern Chinghai, soon after shearing.

tail is large, two lobed, and rather square at the end. It resembles the tail of the Karakul and is more nearly a fat rump than a tail. They are of slightly better mutton conformation than the Mongolian and Tibetan sheep.

Some studies have been made of Chinese and other Asiatic wools and the results of these are summarized below to give an indication of the type and quality of wool produced.

Munz (1937) analyzed several wool samples that originated in northeastern Asia. Four samples of Mongolian wool showed an average of 82.3 percent wool and 17.7 percent hair and kemp. Three samples of Tibetan wool averaged 81.3 percent wool and 18.7 percent hair and kemp. Four samples of Kansu wool, which were similar to the Tibetan

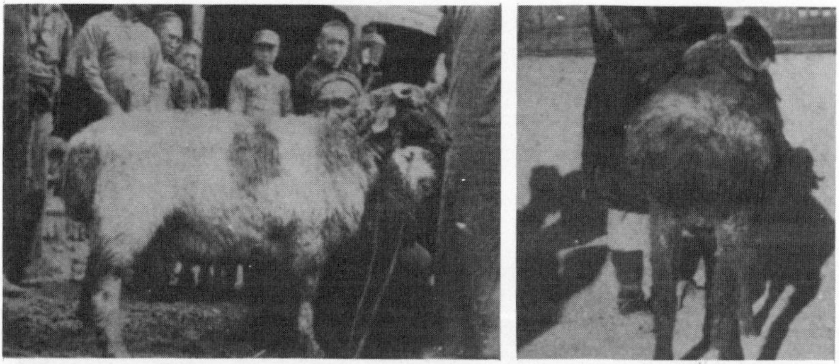

FIGURE 48.—Kazakh sheep. *Left:* A typical animal found in eastern Chinghai. *Right:* Rear view of a Kazakh sheep, showing the fat rump. (Photographs taken by S. F. Wu, Northwestern Agricultural College.)

wool, averaged 82.8 percent wool and 17.2 percent hair and kemp. Two samples from Shensi and Shansi averaged 85.5 percent wool and 14.5 percent hair and kemp. The under wool of both Mongolian and Tibetan types was estimated to have spinning counts of 60 to 70.

Saito (1936) reports that the Mongolian fleeces studied consisted of 22.2 percent coarse wool, including kemp, 8.6 percent semi-coarse type, and 69.2 percent fine type. The average thicknesses of the three types were 90, 42.5, and 25.5 microns.

Burns, Johnston, and Chen (1940) studied a number of samples of Chinese wool and also some other Asiatic wools and wool from the Romney Marsh for comparison. The following types were included in these studies:

1. True Sining wool—a type of Tibetan wool considered to be best of all Chinese carpet wools but rarely obtainable in pure type.
2. Mixed Sining wool—a mixture of true Sining and Kansu wool.
3. Lanchow wool—from Kansu province and often blended with Tibetan wool.
4. Szechwan wool—from Szechwan province and classed with Tibetan wool.
5. Woozie wool—from the neighborhood of Shanghai. Short in staple and suitable for filling wools in carpet-making. Classed among the poorer carpet wools.
6. Vicanere wool—produced in north central India. Very lively and lofty and one of the finest carpet wools.
7. Aleppo wool—grown in Syria and Asia Minor. Has good length, color, strength, and resiliency and is one of the better carpet wools.
8. New Zealand Romney wool—similar to carpet wools but more uniform and lacking the hair and kemp common to carpet wools.

A summary of certain of the data on these wools is given below:

Wool Type	Fiber Type		
	True Wool	Heterotype	Kemp
	Percent	*Percent*	*Percent*
Romney	100.0	. . . . .	. . . . .
Lanchow	91.03	5.83	3.08
True Sining	88.65	9.15	(?) [a] 2.20
Mixed Sining	84.66	3.93	11.35
Woozie	83.72	. . . . .	16.27
Aleppo	74.69	18.87	2.86
Vicanere	59.23	40.67	0.11
Szechwan	87.12	5.80	7.07

[a] (?) refers to an apparent error in original published data.

Wool Type	Fiber Thickness		Stretched Length	
	Mean	Coefficient of Variation	Mean	Coefficient of Variation
Romney	31.99	18.91	218.70	21.32
Aleppo	23.96	24.50	137.30	21.27
Szechwan	20.64	19.70	120.63	35.42
Lanchow	19.56	44.85	107.83	37.00
True Sining	17.51	22.79	110.00	30.26
Mixed Sining	17.43	26.97	79.40	45.04
Vicanere	17.41	19.85	72.00	23.89
Woozie	16.29	21.22	71.00	44.37

Ross (1936) states that the wool of sheep from north Hsingan (Barga region) of Manchuria is very coarse, contains much kemp and hair, and is purely a carpet wool. Wool from South Manchuria (Chinchow and Chaoyang) is described as being finer and containing less kemp. Wool from Hata in central Jehol is described as being intermediate to these two types. The whole fleece of Inner Mongolian sheep is described as having a soft under coat of fine wool and a very coarse outer coat containing much kemp. Ross states that the only wool exported for clothing purposes is obtained by manually combing the living sheep in the late spring when, owing to the fact that the short under coat is shed earlier than the outer coat, a great deal of the under coat may be removed, mixed with only a small portion of the outer coat and containing a minimum of kemp. This wool, classified in the trade as "Sikow ball wool no. 1", was exported to Germany to the extent of 800,000 pounds per year for the manufacture of blankets and coarse tweeds. This type of wool came from southern

FIGURE 49.—Shepherds of northern China. ***Left:*** **A Mongolian herder holding a felt bag used for receiving newly born lambs.** ***Right:*** **Herders in eastern Kansu. Each is equipped with a staff with a mud-throwing spade at one end and a whip at the other, for use in guiding and driving sheep.**

Inner Mongolia, owing to the proximity of Chinese agricultural areas which insured a plentiful supply of labor for hand-combing, which would not be undertaken by Mongols in purely pastoral pursuits.

Ross (1936) mentions that the only other wool produced in Eastern Asia and exported for clothing purposes is the Shantung lamb's wool. This is shorn from lambs at 3 to 4 months and in diameter is equivalent to Merino wool of about 64 spinning count. It is described as uniform and of high spinning value. The wool contains very little kemp or hair at this age but rapidly coarsens in older sheep.

Management and Feeding

Several systems of handling sheep are practiced in China. In the grasslands the animals are grazed throughout the year. During the summer they graze at high elevations to utilize the grass in those areas and to avoid heat and flies. During the autumn, early winter, and spring they are kept on the lower hillsides and higher valleys. During the severe winter weather the sheep are kept in the lower valleys, and at this season some shelter may be provided to protect them from storms. In many areas the sheep may be run in herds along with cattle and horses, one person (usually older men or women, or children) watching the entire group. A sling shot, or other device for throwing stones or soil, is sometimes used as an aid in guiding and controlling the flock. Flocks belonging to lamaseries or to villagers are usually brought inside the walls of the lamasery or village at night, when grazing in the vicinity.

In the farming areas the number of sheep per farm is smaller than the number maintained by nomad families. This number is frequently as small as 1 or 2 in the rice regions and may be as high as 20 to 40, according to Buck (1937). In the northern portions of the farming area the number is higher and some farms keep as many as 100, although most flocks are smaller than this. These animals are grazed on the uncultivated areas on the hillsides and then returned to the farmer's quarters at night. They are also grazed on any uncultivated land on the farm and in the fields after harvest. Overgrazing of the hillsides in such areas is very common. During the winter, straw is fed and other roughage produced on the farm, such as sweet-potato vines, bean vines, etc.

Herding practices vary in the farming areas. In some cases, a child or an old person in the family takes care of the flock. In others, a system of community grazing is practiced. The usual practice is to hire a shepherd to care for all the sheep in a village or belonging to a group of farmers, and he is paid according to the number of animals each has. He cares for them during the day but in the evening each farmer gets his animals, or herding may be rotated among members of the families of the owners. Where they do not have a system of community grazing, farmers or villagers who own only one or two sheep may tether them for grazing or have a child lead them while grazing during the day.

In the area around Tai Lake, west of Shanghai, the common practice is to keep sheep confined throughout the year. This is a silk-producing area and the primary feed is surplus mulberry leaves. The sheep in turn provide meat and wool, and manure for the mulberry trees.

Shearing practices vary in different areas. In Tibet and adjacent areas shearing is usually done once a year, about May, June, and July.

FIGURE 50.—*Left:* Tibetan sheep are sometimes milked. These are tied up for milking on high grasslands east of Lihua. *Right:* Tibetan mastiffs are used for sheepherding and as watch dogs. They are quite ferocious. (Photograph by S. F. Wu, Northwestern Agricultural College.)

FIGURE 51.—*Upper left:* A sheep tied for shearing in Sikong. *Upper right:* Freshly shorn sheep. Shearing is very rough with the crude shears that are available. *Lower left:* A bundle of wool as it usually comes from the Tibetan producers. It is held together by a rope braided from locks of wool. *Lower right:* In some areas, wool is carried part of the way to market on backs of porters, many of them being Tibetan women. Loads are extremely heavy.

In other areas where there is a demand for short wool for use in carpet manufactures, such as in Ningsia, or where the economic pressure makes it necessary to have cash income at more than one season, sheep are frequently shorn twice yearly, in the spring and again in the autumn. Occasionally, they may be shorn as many as three times in a year. In Shansi and adjoining areas of Inner Mongolia, the fine under coat is combed out with a coarse comb in the spring and shearing is done in July, the clip then consisting mainly of outer coat.

Shearing is done with hand shears which are of very poor quality, so that the clip is uneven and the work laborious. Shearers usually do 10 to 15 animals a day. There are reports of men doing as many as 100 per day, but such a number seems highly improbable with the poor equipment that is generally available. In Tibetan regions the shearer usually gets every tenth fleece as pay for the shearing work. The legs of the sheep are usually tied together for shearing. Shears vary in type. Some have a spring-type hinge while others have a bolt pivot hinge, and a few are still in use that are in two parts with the hinge provided by the shearer's hand. Providing of good-type shears, made of high quality steel, would be a big contribution to the sheep-raisers and also the users of wool since they would be supplied with a product of more uniform length. Often, in the Tibetan area, the loose wool is hand-pulled before shearing is done. Women and children are on the alert, during the spring season, to pick up wool that has been shed.

Docking is seldom practiced. In the Tibetan sheep it is hardly necessary, owing to the small size of the tail. In the other types which have fat tails, it is rather difficult but might not be entirely impractical if done at an early age. Research is needed on this point.

Ross (1936) describes in some detail the feeding and other environmental conditions for sheep in Manchuria and adjacent regions. Only a few points from his discussion will be mentioned here. He points out that in Manchuria (except in the extreme south) and on the Mongolian plateau, little growth of plants occurs before the end of April and after August, with the result that for six or seven months (eight in the case of Hailar) sheep are required to subsist on dried summer grasses. The effect of this is accentuated over a large part of the Barga plain and on the Mongolian plateau by the complete lack of cover on the bare and treeless prairie. Ross also states that normally little snow falls in this region owing to the extreme dryness during the winter months, and native sheep are able by scratching to reach the dry grass underneath the snow. If, however, a snowfall is unusually heavy, even of only a few extra inches, then high mortality may result, as during 1933–34 when the wool from the Barga region decreased by 25 percent, largely owing to mortality from starvation. In another area, Darahan in South Hsingan, Ross cites the great difficulty under existing conditions for growing improved sheep, there being no growth of grass from October to May and no winter-growing legumes. Heavy rain in June and July is reported to cause rank growth of coarse reed-like grass which is of little use for grazing or haymaking. It is also stated that the sheep in this area must be housed at night for protection against both bandits and wolves.

Sheep are utilized for mutton almost entirely in and near the areas where they are produced, since there are no facilities for moving animals for long distances for utilization of the meat in other areas. Herds for slaughter move on foot from grassland areas to such cities as Peking and Tientsin that are within driving distance. Therefore, the use of mutton as a staple article of diet is limited to the grasslands, cities in and near the grasslands, and to farming areas where a con-

FIGURE 52.—Sheep are important to the people of the grasslands. *Upper left:* A Tibetan man wears a robe lined with sheepskin. The right shoulder is usually exposed during the warmer portion of the day. *Upper right:* Mongolian and other groups use woolen felt to cover their yurts or movable houses. This one belonged to Kazakh tribespeople that had migrated from Sinkiang to eastern Chinghai. *Lower left:* Mutton is the chief meat eaten in the grassland areas. The cuts in the fat are made so that the customer may see the lean meat. *Lower right:* Mohammedan soldiers of the Chinghai provincial army make many of their own clothes. Here they spin yarn on the typical native spindle.

siderable number of sheep are raised. Many animals are slaughtered and utilized by the producers. When sold in villages and cities, the carcasses are hung up in the small open-air markets and pieces cut off when a purchaser arrives. In the northwest it is common practice to cut the fat in several places, either by a cross or by a series of parallel cuts, to expose the lean so the customer can see it. This is the opposite of practices in other countries, where the fat cover is preserved as carefully as possible to protect the lean and also to show that the animal was well fattened.

Pelts and skins are used for several purposes. The pelts, particularly of lambs and of younger sheep, are valued for use in lining gowns for winter wear. Skins are used for leather for various purposes, and

FIGURE 53.—Rugmaking is an important home industry in the grasslands and has been commercialized to some extent. *Top:* Tibetan women sorting wool, spinning, and weaving. *Middle:* A rug factory in Gyantse, Tibet. *Bottom:* Ningsia rugs. Many of the rugs woven in this province are of excellent quality, design, and color.

a great number are used in sheepskin rafts. Skins for this latter purpose are soaked in water for a few days, and then the wool is pulled out. Linseed oil is then placed inside, shaken around, and allowed to soak in thoroughly. The skins for this purpose are removed carefully so that the leg, neck, and rear openings are as small as possible. After the soaking with linseed oil, these openings are carefully sewed up, and the skin is inflated with air. The inflated skins are fastened to light wooden frames and used as rafts in river transport. The small rafts have 10 to 13 skins and usually 12 or 13. After a trip downstream, the operator picks the raft up and walks back upstream for the next load. Small rafts may be fastened together for larger loads. For long trips down the river, as many as 300 skins are made up into a single raft.

FIGURE 54.—Buoyant sheep- and goat-skin rafts are used to carry passengers and goods down and across the Yellow River. Small rafts contain 11 to 13 inflated skins. Larger ones, used in long journeys, contain up to 300 skins. *Left:* A raft is carried back upstream on a coolie's back. The current is too swift for paddling these rafts upstream. *Right:* Two small rafts are tied together to ferry passengers across the river.

The Wool Trade

Wool is the chief commercial commodity resulting from sheep production in China. It is of very poor quality from the standpoint of clothing manufacture and has been utilized to a large extent in world trade as carpet wool. A considerable amount is also utilized by natives in the grassland-producing areas and areas nearby for spinning, felting, and weaving into rugs, felt mats, and to some extent for clothing. During the present war a considerable amount is being used in commercial textile factories.

Some wool is transported to market by truck, but most of the transporting is still done by pack animals and by rafts and other river

615999—45——7

craft. Carts are also used in some areas. Shearing equipment and methods are poor, so the wool is generally in poor condition when it is ready to be sent to market. It is usually made up into bundles, weighing 80 to 140 catties, which are tied with rope that is braided from wool. The identity of individual fleeces is completely lost. Adulteration of wool with dirt and other materials, either by the producer or by intermediate purchasing agents, is very common. Because of these practices and lack of proper precautions to keep the wool dry, it usually reaches the central marketing points in extremely poor condition. At these central marketing points the wool is cleaned up by dusting and sorting out the dirt. It may then be sorted into two or three grades, largely on the basis of relative fineness, color, et cetera. No well-organized methods of grading or sorting have been developed.

Wool is handled as a government monopoly in some provinces, and the wool of the northwest is to a considerable extent being handled as a central government monopoly, through a government-buying agency in Lanchow which buys from the producer and resells to the manufacturer or for export. There is much room for improvement in the method of handling wool since there is at present much wastage and inefficiency. Some of the wool is washed before reshipment to the manufacturer; this is being done in Lanchow by an establishment that is operated jointly by the central government and a commercial textile concern.

The amounts of wool exported from China in various years are shown in the table below. Goat's hair and wool are also included in this table for convenience.

Year	Goat's Hair (quintals [a])	Goat's Wool (quintals)	Wool (quintals)	Total (quintals)	Total Value of Exports (percent)
1915	----------	6, 210	228, 517	234, 727	2. 76
1920	----------	5, 052	62, 747	67, 799	. 61
1925	14, 173	14, 068	257, 807	286, 048	2. 11
1930	5, 486	7, 519	118, 212	131, 217	. 42
1935	13, 591	11, 949	199, 871	225, 411	2. 81
1940	7, 123	8, 280	9, 517	24, 920	. 76

[a] Quintal=100 kilos.

The figures for the percentage of total exports that consist of wool, in terms of monetary value, indicate that wool does not make a large contribution to the total. However, this income is very important to a section of the population which depends largely on livestock for its existence.

The countries to which most of the wool has been exported in recent years are indicated below, in percents of the total wool going to each:

Year	United States	Great Britain	Germany	Japan	Others
1930	73. 33	5. 94	2. 05	14. 65	4. 03
1935	83. 77	3. 30	6. 57	3. 72	2. 64
1940	54. 03	24. 85	. 30	14. 58	6. 24

The relative importance of various exporting points is indicated by the following figures, which show the percentage of the wool that was exported from various cities in recent years:

Year	Tientsin	Shanghai	Chungking	Mengtze	Others
1930	70. 89	5. 03	12. 14	. 03	11. 91
1935	85. 59	14. 40	..	. 01	..
1940	32. 73	65. 45	..	. 87	. 95

Exports have naturally decreased owing to wartime conditions, and this is especially true of exports from seaports. During the present war, Lanchow has become the most important central marketing point for wool. Some wool from Tibet goes on pack animals into India.

Much of the wool exported now goes to the Soviet Union, and the estimated amounts exported to that country in 1938 and 1939 were 17,000 and 530,000 piculs,[1] respectively.

A considerable amount is imported into China. The amount in 1937 was 16,110,465 pounds. Of this, 11,573,354 pounds were from England, largely in the form of wool top. Other imports included 1,178,212 pounds of yarn and 70,850 square yards or 2,376,567 pounds of woolen piece goods.

Much of the wool manufacturing in China at present is hand work, either for home use for clothing or for local sale. There is also some rug making in some areas, especially Ningsia, Kansu, Sikong, and Chinghai. The manufacture of felt, mostly by individual or small manufacturing units, is also important in these and other areas. A few large plants are operating in and near larger cities for the production of various types of woolen cloth. Industrial cooperatives are doing some wool-manufacturing work, especially for army blankets.

Prior to the present war there was a marked increase in woolen manufacturing in China and this will undoubtedly be resumed when

[1] 1 picul=60.5 kilograms.

peace comes. Ross (1936) states, for example, that 2,000,000 pounds of native wool were being used for clothing purposes in Mukden. It was being used to make cheap blankets for the local Chinese demand and being mixed with a small portion of imported wool for army blankets. Uniforms for the Manchu army and police were being made with a warp of native wool and a weft of imported wool. The growth of woolen mills in various Chinese cities is also cited. Prior to 1934 Chinese imports consisted very largely of manufactured goods with only a small amount of tops and an even smaller quantity of grease wool. In 1934 and 1935 there were marked increases in the volume of tops imported and a sharp decrease in the imports of woolen goods, especially yarns. Ross points out that most of the tops used are of the coarser type (48 spinning count), coarser yarns being used for the production of low-priced fingering wools which are used by the Chinese lower classes for the knitting of mufflers and caps, while a limited amount of higher-quality tops was being utilized for production of worsted materials, gabardine and serges, etc., for the higher-class Chinese demand. Ross states that there is widespread opinion that there is a definite transition in China from the use of cotton and silk to the use of wool. Price is indicated to be one of the main limiting factors in this transition, and ignorance of the value and use of wool among the illiterate population is a second limiting factor.

Breeding

Almost no data are available on the basic reproductive phenomena of the age of puberty, length of estrus and estrous cycle, and duration of pregnancy.

In the range-sheep areas, the ewes apparently begin coming in estrus about the first of August, and most of them are bred during August, September, and October. This results in lambings in winter and early spring when the weather conditions are unfavorable to survival and feed conditions are not satisfactory for a good flow of milk from the dam. Very few attempts are made to control the breeding season, the rams being allowed to run with the ewes during the entire year. It is reported that a birch-bark apron is used in some sections of Sikong in an attempt to delay breeding. It is reported that most of the ewes in Yunnan are bred during the summer, so that lambs are born in the fall and early winter. Losses are said to be heavy owing to this lambing at an unfavorable season. Work is needed to determine the length of the breeding season in the various grassland areas of China, as a basis for determining if breeding can be delayed and yet get a satisfactory lamb crop. Owing to poor feed supply during late winter and early spring it is possible that ewes may be in too poor condition to carry and nurse lambs, if bred in late fall or early winter. Investigations are needed on this point.

Observations have been made during 1942 and 1943 by Hsu and Peng (1943) which shed some light on this problem. A group of Tibetan ewes were purchased in 1942 for experimental breeding work and most of them were already bred. Of the 16 that were not bred, 7 were settled in July, 2 in August, 3 in September, 5 in October, and 1 in December of 1942. In 1943, breeding was started during the latter part of June. Six ewes were bred in June, 34 in July, 21 in August, 5 in September, and 3 during the early part of October. Breeding was not yet completed at the time of preparation of this report. However, these data indicate that the Tibetan sheep tend to have a prolonged breeding season, and if they follow the pattern of other types of sheep, it should be possible to postpone breeding until the early winter months so that lambs would be born after the weather warms up and the grass begins to grow in the spring.

Duration of pregnancy of the 16 ewes mentioned in the preceding paragraph as being bred in 1942 averaged 152.6 days and ranged from 148 to 159 days. Birth weight of the lambs averaged 5.65 catties.

The situation differs somewhat in the farming areas from that in the grasslands. In Shansi, for example, ewes usually begin to come into heat at about the time summer rains bring out a growth of fresh grass, ordinarily in the first half of July. A few ewes come into heat in June. Shepherds prefer to breed as early in the season as possible. Ewes remain in fairly good condition through January or February. Subsequently, because the pasture is extremely scanty they are in condition neither to nurture a foetus or suckle a lamb. Lambs born in March or later are often stunted at birth and develop poorly afterward. Temperature conditions are less favorable for survival of the lamb when born in the middle of winter, but provision is made to take care of that in areas of settled agriculture like Shansi or Suiyuan. During lambing season shepherds are on hand night and day to care for ewe and lamb. Following birth, both ewe and lamb are kept in the farm enclosure for several days. After the ewe is returned to the flock and pasture, lambs are kept at home under shelter until growth and advancing spring allow greater exposure. Therefore, loss of lambs from exposure in these areas is not high. The long coat of wool carried by lambs at birth may increase their ability to stand cold. In Suiyuan, shepherds often collect lambs born on the range and place them in pockets of heavy felt bags, in which they are carried back to shelter.

Sheep in Shantung are reported to be prolific, producing many twins, but definite data are lacking and, in general, twinning is rare among Mongolian sheep. Lamb-crop figures are not available for the grassland areas, but some estimates place the crop at 50 to 60 percent born. Death losses are usually heavy, the extent of these losses depending upon weather and feed conditions, and estimates obtained of the losses are sometimes as high as 50 to 90 percent. The liver fluke, which is present in many grassland areas and especially the damper regions of the Tibetan country, takes a heavy toll of lambs.

Most males are castrated. Several techniques are used. The scrotum may be cut with either scissors or knife, the testes squeezed out, and the cord severed by cutting, by pulling with hands or teeth, or by biting it. In some sections a string may be tied around the base of the scrotum in young lambs. The practice of crushing the spermatic cord by beating between two stones is also reported.

The proportion of males left intact varies in different areas. In the grasslands of the northwest the common practice is to leave 1 intact male for every 10 females. Where large herds are maintained the wethers are usually kept in one herd and the ewes and rams in another. In Sikong there are usually 5 rams per 100 ewes, but the smaller flocks usually have a higher proportion of rams.

Chang (1943) made a comparison of the fleeces of 18 Tibetan (classified as valley type) sheep and 4 animals from Tibetan ewes and sired by a Rambouillet ram. These data are summarized below:

Items Compared	Tibetan	Rambouillet × Tibetan
	Catties	*Catties*
Fleece weight	1. 5	5. 0
Percent water	12. 75	17. 09
Percent yolk	9. 23	14. 50
Percent impurities	20. 0	19. 0
	Percent	*Percent*
True wool	72. 71	79. 68
Heterotype	13. 16	18. 66
Kemp	14. 13	1. 66
	Length (*cm.*)	*Length* (*cm.*)
True wool	16. 63	12. 75
Heterotype	18. 70	15. 10
Kemp	6. 63	. 56
	Fineness (*1/10,000 in.*)	*Fineness* (*1/10,000 in.*)
True wool	7. 54	8. 03
Heterotype	2. 56	12. 36
Kemp	12. 99	11. 13

The individual fiber data represent averages of 300 fibers, 100 taken from each shoulder, side and thigh of each animal.

Kislovsky (1938) states that 1,300 Merino rams had been brought into Outer Mongolia (apparently from the Soviet Union) for use in grading up the native sheep but no details are given of the results obtained.

A considerable number of Rambouillets have been brought into Sinkiang from the Soviet Union, but no detailed data are available on results obtained from their use in grading up. An indication of the result obtained is given in figure 59.

Data on the results of grading up Mongolian sheep with Merinos [2] in Manchuria, as measured by fleece yields and reviewed by Wang (1942), are listed below:

Breeding	Males		Females	
	Number	Fleece Wt.	Number	Fleece Wt.
		(grams)		*(grams)*
Mongolian	356	1, 610	2, 928	1, 182
Mongolian × Merino	356	3, 421	2, 928	2, 456
Improved Mongolian × Merino	356	4, 581	2, 928	2, 863
Merino	193	8, 060	427	5, 860

The exact breeding of the "Improved Mongolian×Merino" is not clear. It is stated in this same report that the South Manchuria Railway Company began work with Merino and Corriedales in 1934, but no results are given.

Ross (1936) visited Manchuria and adjacent areas in 1936 and observed the results of several attempts to use improved breeds for grading up of the native sheep. Some of his observations are summarized briefly below.

A considerable amount of work has been done at the Kunchuling station, maintained by the South Manchuria Railway Company, near Hsinking in Kirin Province. This station is not in a pastoral area so the results cannot be taken to indicate performance of improved sheep in the native sheep-raising areas. First-generation sheep from the use of Rambouillet rams showed a considerable increase in wool production, but the wool still had much of the character of the native wool, being fluffy and lacking density while still being mixed with a considerable portion of the coarse fibers of native sheep. The amount of wool is about double that of the native or about four to five pounds. The second-generation sheep (¾ Rambouillet) approached Merino wool type, but the staple was somewhat longer and density was less, and there was little wool on the neck, legs, and belly. There was little evidence of native-type wool fibers or hair. These sheep were housed and well bedded and were fed what is described as a generous ration, but the condition of the sheep and of the wool suggested undernutrition. Corriedales and first-generation grades of Corriedale and

[2] These may have been Rambouillets rather than Merinos, for the two names are often confused in China.

Mongolian sheep at this station were described as small and undersized, and as having wool that lacked density and character.

Work conducted at the Darahan station in South Hsingan gave a better indication of the ability of improved types to survive under native conditions, since this station was located in the border country between farming and pastoral regions. First, second, and inbred cross Merino-Mongolian sheep were all available at this station. They were housed at night for protection against wolves and bandits but otherwise were handled under natural grazing conditions as far as possible. Purebred Merinos appeared to show the effects of the severe winter and unsuitable pasture conditions in the summer. The staple was very short and rather oily and contained much dirt, yielding only 30 percent after scouring. The sheep were in poor condition and rams yielded only 8 to 9 pounds, rarely 10 pounds, and ewes 6 to 7 pounds. First- and second-cross and inbred second-cross sheep showed the expected increase in wool yield but a marked decrease in clean scoured yields, from 70 percent in the native to 60 and 40 in the first and second crosses. First-cross animals appeared to be in better condition than the Merinos or the second-cross animals. All the improved animals were fed a staple ration of soybean straw, supplemented with some lucerne hay and grain (principally kaoliang) during the winter months. The director of the station stated that he did not believe crossbred sheep could survive under natural conditions during the winter, owing to less resistance than the native sheep and relatively little aptitude for scratching through the snow to find feed.

A considerable number of Merinos were imported into Shansi Province by the Governor, Yen Shi Shan, about 1919 or soon thereafter. Ross (1936) states that these were kept under controlled conditions, being hand-fed and housed, and increased up to about 2,000 in number. When an attempt was made to run them under natural conditions, with only some feeding of hay in the winter when snow was heavy, the flock deteriorated rapidly and numbers declined owing to mortality. Ross also states that attempts at crossbreeding (grading up) in the hope of increasing the hardiness of the sheep are said to have failed, there being little increase in their resistance to adverse factors, while the wool was of low value and very coarse and hairy. The work in Shansi was apparently done with no adequate technical supervision, hence it is difficult to evaluate the results. Ross inspected a small flock in the Saratsie District of Suiyuan which originated from these Shansi Merinos. One ram and 10 ewes had been run with some native ewes for three years. This area is about 2,000 feet lower than the Mongolian plateau and considerably warmer. They had apparently thrived in this region but when moved to the plateau death losses were heavy. At the time of inspection, only one of the ewes that originally came from Shansi remained. This and others of the pure or grade Merino type had wool only 1¼ to 1½ inches in length, although this represented six months' growth during the summer months. All wool of the Merino types was unusually fine, as would be expected under conditions of inadequate nutrition. The grade sheep had a marked admixture of coarse Mongolian-type wool with much kemp and heterotypic fiber. The three-quarter Merino sheep were reported to shear 6 to 7 pounds as compared with 2 to 2½ pounds on the native sheep, which were valued at about 50 and 20 cents per pound, respectively.

Experiments in crossbreeding between the Rambouillet and certain of the types of sheep native to North China were conducted over a period of nine years by Moyer and Tong (1944) at Minghsien College in Taiku, Shansi, to investigate the possibilities of producing finer grades of wool in China by that method. As part of the project, specific attention was also given to a study of the local types themselves. The plan of the experiments and the available results are summarized below.

Four native types were selected, to which the designations Taiku, Shouyang, Hanyang, and Luan were given. The regions where these types are found, as far as known, and certain of their distinguishing characteristics are described below.

The Taiku is the type found most commonly in Shansi, as well as in Hopei, Shensi, and possibly other provinces. It resembles closely the type found in Inner Mongolia, from which it may not differ significantly. In contrast with that type, faces are usually white and the animals appear to be somewhat smaller in size. Its fat tail is relatively small and its wool covering contains a finer under coat and a coarser outer coat.

The Shouyang type is found as a distinct type in Shouyang district, Shansi, and to a lesser extent in surrounding districts where it often occurs mixed with the Taiku. It differs from the Taiku in having a heavier tail and shorter legs, in being somewhat lighter in weight, and in having a heavier fleece which is somewhat wavy and lustrous. Its wool is recognized as a distinct class on the Tientsin market.

The Hanyang type is found in western Honan, particularly in the vicinity of Hanyang city. A similar animal, found eastward in Shantung, is believed to be of the same type, and that found around Tungchow, Shensi, described elsewhere in this publication seems to be similar. It is characterized by an unusually heavy fat tail and by a finer wool which also is somewhat crimpy. Its wool is relatively uniform in character, without a distinct inner and outer coat. On the Tientsin market this wool is considered as the finest of the Chinese wools. Unlike the Taiku and the Shouyang, twinning is fairly common in this type.

The Luan is a name given to a type after the former name of a city in southeast Shansi (now called Changchih) around which the type is commonly found. It is distinguished by its tail, which is long like most western breeds but is enlarged around the base, resembling the

FIGURE 55.—Results of grading up Mongolian sheep of the Taiku, Shansi area, with Rambouillet rams at Minghsien College. ***Top, left to right:*** **Taiku ewe; ½ Taiku – ½ Rambouillet; ¼ Taiku – ¾ Rambouillet; and Rambouillet.** ***Middle:*** **Group of young ½ Rambouillet – ½ Taiku rams.** ***Bottom:*** **Group of ¾ Rambouillet – ¼ Taiku ewes.**

FIGURE 56.—A study of variation in tails. ***Left to right:*** **Rambouillet (docked); ¾ Rambouillet - ¼ Shouyang; ½ Rambouillet - ½ Shouyang; Hanyang; Shouyang; and ½ Rambouillet - ½ Hanyang.**

tail of first-generation Rambouillet-Taiku crossbreds. In size and in the characteristics of its wool it resembles the Hanyang type.

The Rambouillets used were mostly of a smooth-bodied type, obtained from the Deer Lodge Farms Company, Montana, although some lightly wrinkled rams, imported from Wyoming and Montana by the Chinese Government, were also used. The Rambouillets included both rams and ewes, and their offspring were utilized in a study of the behavior of this stock under local conditions.

The first experiments, begun in 1932, included only two breeds, the Taiku and the Shouyang. Between 30 and 40 ewes of each of these types were successfully bred, approximately one half being with rams of the Rambouillet and one half with rams of the appropriate native type. Later, with help from the National Economic Council of the Chinese Government and from the Rockefeller Foundation, the experiment was very much enlarged. Two additional types were introduced, the Hanyang and the Luan, and the number of ewes of each type in the experiment was increased to about 120. A part of this increase in program was accomplished in 1936 and the rest was completed in 1937. The Japanese invasion, however, completely ruined the work of that year, and many sheep were lost. A portion of the program was reestablished in 1938 and continued into 1941.

Crossbreeding, in the main part of the program, was not carried beyond the first and second crosses with the Rambouillets, which produced half-blood and three-quarter-blood Rambouillet animals. It was considered that little additional information of value would be gained by going further. During the last several years of the experiment, however, a large number of Rambouillet-Taiku half-bloods were backcrossed to the Taiku in order to bring out the inheritance of certain genetic factors and in order to investigate the beginning steps in one plan which might be followed in fixing a new improved type.

Experimental animals of all types were raised during any given season under approximately the same environmental conditions.

FIGURE 57.—Results of grading up Taiku sheep with Rambouillet sheep at Minghsien College. *Top, left to right:* Rambouillet ram, Taiku ewe and ½ Rambouillet - ½ Taiku ewe. Of the two piles of wool in front of the Taiku ewe, the left is fine under coat pulled in the spring and the right is coarse outer coat and a little under coat clipped in July. *Bottom:* Pulling wool (under coat) from Mongolian sheep with a coarse comb in the spring.

These were a little better than the conditions under which local animals are commonly raised, but not very much better. Breeding-stock gained the main part of their sustenance the year round by grazing in the surrounding country. A certain amount of supplementary feed-stuffs was given during the poorest months for grazing, usually between January and April, consisting of materials available, such as leaves, sweet-potato vines, buckwheat straw, bean straw, and bean pods. No concentrates were fed, except to the Rambouillet stock which was grazed with the rest but which otherwise received favored treatment. All animals were quartered in the open except during the coldest months, when they were placed at night in simple unfloored sheds. Pens and yards were swept daily, thus reducing to some extent the danger from parasites.

The breeding season, following local practice, began in July and usually ended in October, and the lambing season began in December. Lambs were given careful attention at birth, and following birth they were kept in pens and yards until about April when they were pastured during the daytime. Upon weaning they were fed mainly with carrot tops, the preferred food for lambs in that region. No concentrates were fed.

Weights and measurements of all offspring were taken at birth and at intervals of one month thereafter until reaching an age of 6 months. Similar records were taken subsequently at 9, 12, and 18 months and afterwards twice a year, following the spring and fall shearings. Measurements included length of legs, height at the shoulders, body circumference, width of loins, circumference of tail at base, and others. Weights and similar measurements were taken of the breeding-stock twice yearly, after the spring and fall shearings.

Local animals were sheared according to local practice. For the Taiku and Shouyang types, this consisted of pulling out the finer under coat with a coarse, claw-like comb, in the spring. In the latter part of July or early August the animals were clipped, removing the coarse outer coat and whatever else remained or may have grown. The Hanyang and the Luan types and all crossbreds were sheared once annually, in the spring.

Records kept with respect to wool included weight after shearing and the scoured weight of a representative sample. For scouring, one-pound samples were picked out of different parts of the sheared or pulled wool spread out on a table. Samples were washed several times, first with soda, and then thoroughly rinsed. A special laboratory was set up for this purpose, which included washing apparatus and a drying oven. For future reference and detailed study, shoulder samples were taken from each animal at shearing time and filed.

In all, records of these kinds were obtained from some 400 to 600 animals of native types, and from several hundred crossbred types. Triplicate copies of these data were prepared and left in different places for safekeeping. It is hoped that at least one set will be found intact after the war is over, when the data will be summarized and analyzed.

A detailed statement of the findings of these investigations can be made only after this analysis has been completed. Information is sufficiently complete, however, to state in a general way some of the results obtained. With respect to body weight, first-generation crossbreds of Rambouillets with the Taiku and Shouyang types were appreciably heavier than the native animals. With respect to meat, a local prejudice existed against the flavor of the Merino breed, from experience gained with animals introduced by Governor Yen Hsi-shan. A difference in flavor could usually be detected also in meat of the Rambouillet crossbreds, in comparison with meat of native animals, but objection to it was not strong.

Average wool production of the Taiku type was around 2 pounds. That of the first-generation cross with Rambouillets was about 5 pounds. The three-quarter bloods averaged approximately 8 pounds, and individual animals went up to as high as 10 pounds. The quality of wool also was better. The fleeces of the half-bloods carried a certain amount of a coarser outer coat, but this type of fiber was relatively inconspicuous in the ewes, although more prominent in the rams. Fleeces of the three-quarter-bloods contained no coarse outer coat, and the wool throughout was finer and contained more yolk. Shouyang native animals produced 3 pounds of wool per animal, and the crossbreds were similar to those of the Taiku except that the wool, in general, seemed a little coarser.

The following table shows data available for the year 1935, from experimental offspring which then were 18 months or older. It should be mentioned that wool yields for all types of animals that year were, in general, lower than the average.

Type of Sheep	Number Averaged	Average Body Weight	Average Wool Yield Unscoured	Taiyuan Price per Lb. of Wool
		(Lbs.)	*(Lbs.)*	*(Chinese yuan)*
Taiku	14	71. 3	1. 58	. 15
Taiku-Rambouillet half-blood	19	83. 9	4. 37	. 38
Shouyang	12	66. 0	2. 77	. 18
Shouyang-Rambouillet half-blood	14	87. 7	4. 89	. 38
Rambouillet	3	125. 7	11. 78	. 47

With respect to hardiness it was soon demonstrated that the Rambouillet breed was not as hardy as the native breeds under local environmental conditions. Stomach worms killed all the lambs born the first year, but this pest was fully controlled in subsequent years by drenching with copper sulphate. Half-bloods seemed only slightly less resistant than native animals to this and other adverse factors in the environment, but three-quarter-bloods were definitely less hardy. This was shown in a greater proportion of deaths before reaching maturity and in a less healthy appearance of those that survived. Experience in these studies suggests that no animal less hardy than the half-bloods would be raised successfully in that region unless the present environmental conditions were appreciably improved.

Available information is as yet insufficient to report on results obtained with the Hanyang and Luan types.

An attempt was made to study the inheritance of the tail characteristics. Length appeared without exception to be dominant. All first-generation crossbreds had a long tail, resembling the Rambouillet, although the comparative number of vertebrae was not determined. The curl sometimes found at the end of a fat tail seems to be inherited as a separate character, possibly as a single factor. Some long tails of the crossbreds had a slight twist at the bottom, while others were without it. Fattiness appears to be governed by multiple factors. First-generation crossbreds had a partial fattiness, which expressed itself as an enlargement around the base. In the case of the Rambouillet-Hanyang half-bloods, the amount of fat was quite considerable. Three-quarter Rambouillet–one-quarter Taiku animals had little or no enlargement of the tail.

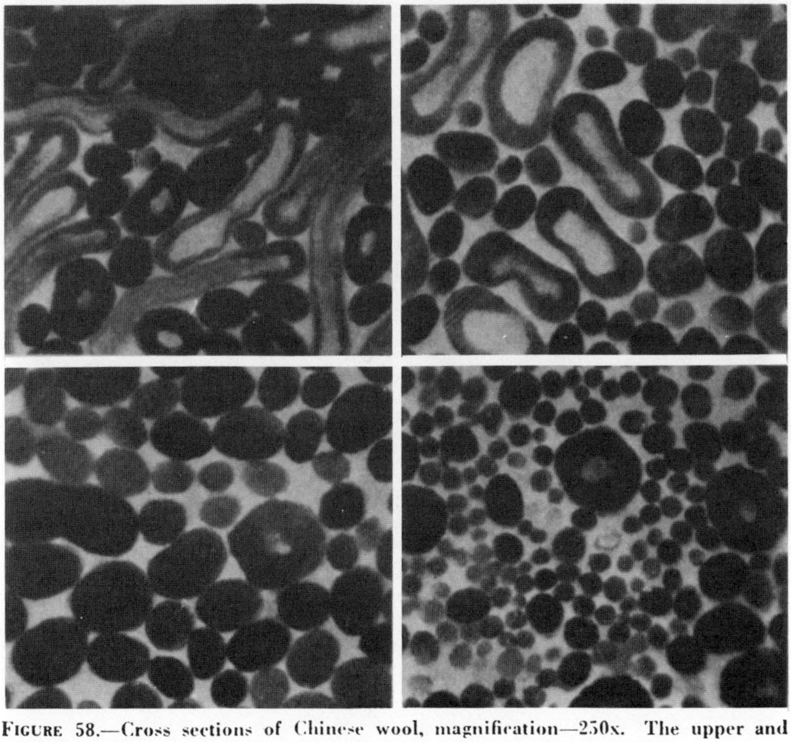

FIGURE 58.—Cross sections of Chinese wool, magnification—250x. The upper and the lower-left pictures are of samples from mature Tibetan wethers in eastern Chinghai. The lower-right picture is from a mature Tung sheep in Shensi.

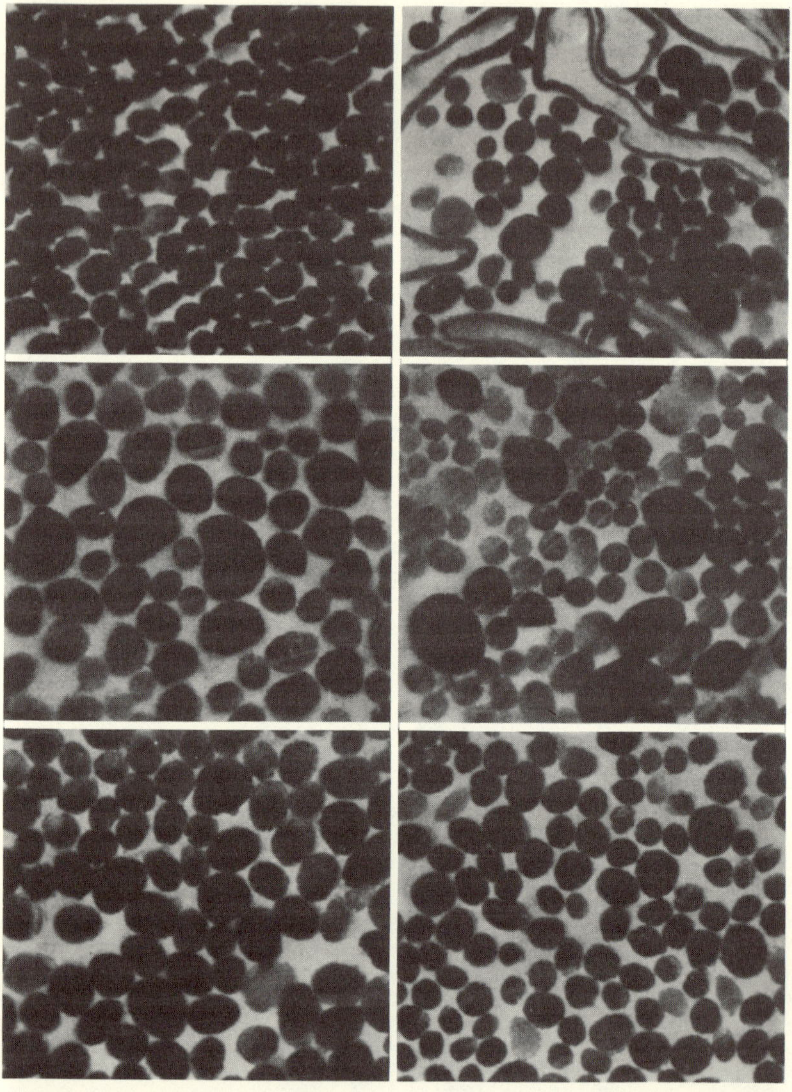

FIGURE 59.—Results of grading up Kazakh sheep with Rambouillet rams, in fineness of fiber. Photographs of cross-sections are of side samples at 250x. *Upper left:* Rambouillet ram, age 5 years. *Upper right:* Kazakh ewe, age 2 years. *Middle left:* ½ Rambouillet – ½ Kazakh ewe, age 2 years. *Middle right:* ¾ Rambouillet – ¼ Kazakh ewe, age 2 years. *Lower left:* ⅞ Rambouillet – ⅛ Kazakh ewe, age 2 years. *Lower right:* 15/16 Rambouillet – 1/16 Kazakh ewe, age 2 years. The yields of grease wools of Kazakh and the four successive grade ewes were 1.5, 3, 4, 5, and 7 catties, respectively. The original wool samples and the data were furnished by K. T. Hsu, Chinese Ministry of Agriculture.

Goats

Types

The goats of China are quite variable in size, type of coat, and color. No distinct types are recognized over large areas, but there are groups in many localities that appear rather distinct.

Colors that are commonly seen are solid black, tan with a dark stripe down the back and some white on the face and underline (somewhat like the Toggenburg pattern); white; gray or blue (white and black fibers mixed); black and white (black may be on front and rear portions); and dark fawn or brown with a dark stripe down the back. Kislovsky (1938) reports that gray tones predominate in goats in Outer Mongolia, the proportion being 66.8 percent.

Both polled and horned animals are found, and there is much variation in size and shape of horns. There is also wide variation in the length of hair. Many animals have short, smooth coats while others have quite long, heavy coats of hair, and there are all gradations in between.

The "Mah" goat is a rather distinct type found in the vicinity of Chengtu, Szechwan. Color ranges from light to deep brown. A dark brown or black stripe extends down the face, with lighter brown on either side, and a dark brown stripe extends along the back from horns to tail. Darker hairs are found along the midlines of the fronts of the legs. Both males and females are horned. A few polled animals are found. In males the horns are short and thick, the tips pointing outward and downward, while in the female they are more slender and point upward and backward. Ears are straight and point upward and forward. The back is narrow, the abdomen is broad and deep, and the legs are relatively short. Hoofs are usually black. Hair is short over most of the body, but in some animals it may be rather long on the upper portions of the legs and along the sides of the abdomen.

Two types of udders are found in the "Mah" goat. A cylindrical type, with large teats which point downward and forward, is found in the larger-type animals, while smaller animals tend to have spher-

615999—45——8

ical udders with small teats that point somewhat to the side. Forty-nine adult females were measured by Tang (1943) and these data are summarized below:

Item	Average	Range
	Cm.	*Cm.*
Length, shoulder point to pinbone	70. 7	61–82
Height at withers	63. 5	51–71
	Catties	*Catties*
Weight	73. 7	50–98

Six adult males were also weighed, which ranged from 76 to 101 catties and averaged 84.8. In disposition, the "Mah" goats are active and rather bad-tempered. They are good grazers. The lactation period averages about 5 months but may be as long as 10 months. Data obtained on milk samples from 23 does, apparently at varying times in lactation, showed the following averages: specific gravity, 1.032; fat, 6.8 percent; milk sugar, 3.38 percent; milk protein, 5.13 percent. Data on 20 does showed average production to be 267.7 c.c. per day, with 466.7 c.c. as the highest observed value. These goats are used chiefly as a source of meat and the hides are used for leather. Most of the hides were exported prior to the war. Since the farmers usually do not milk their does, little has been done to improve milk production in the native animals.

A small black or very dark-brown type with long hair is found in Shensi and Shansi. It is said to produce up to three pounds of milk per day at the height of lactation. The extent of the area in which it is found and the variations in size and other characteristics in this type are not known.

In Yunnan a type of black goat is said to exist which is horned, muscular, fattens reasonably well, and has an adult weight of 60 to 70 catties (30 to 35 kilos). It is reported to produce about 1 catty (½ kilo) of milk a day.

Gorosenko (1936) states that all goats throughout Mongolia belong to one primitive type. The color is variable, but 66.8 percent are gray. The live weight is 30 to 41 kilograms, and the height at withers is 59 to 63 cm. The majority, 73.4 percent, are horned. Gorosenko believes that all goats of Central Asia (Kirghizstan, Kazakhstan, Tadjikistan, and Mongolia) are of a common type. He states that goats in Mongolia are used for milk, meat, and skins.

Other local areas no doubt have predominant types but there is not sufficient information available to indicate either the types or the distribution in many areas. In many sections there is a multiplicity of types. In Chinghai, for example, a great variety of colors are seen

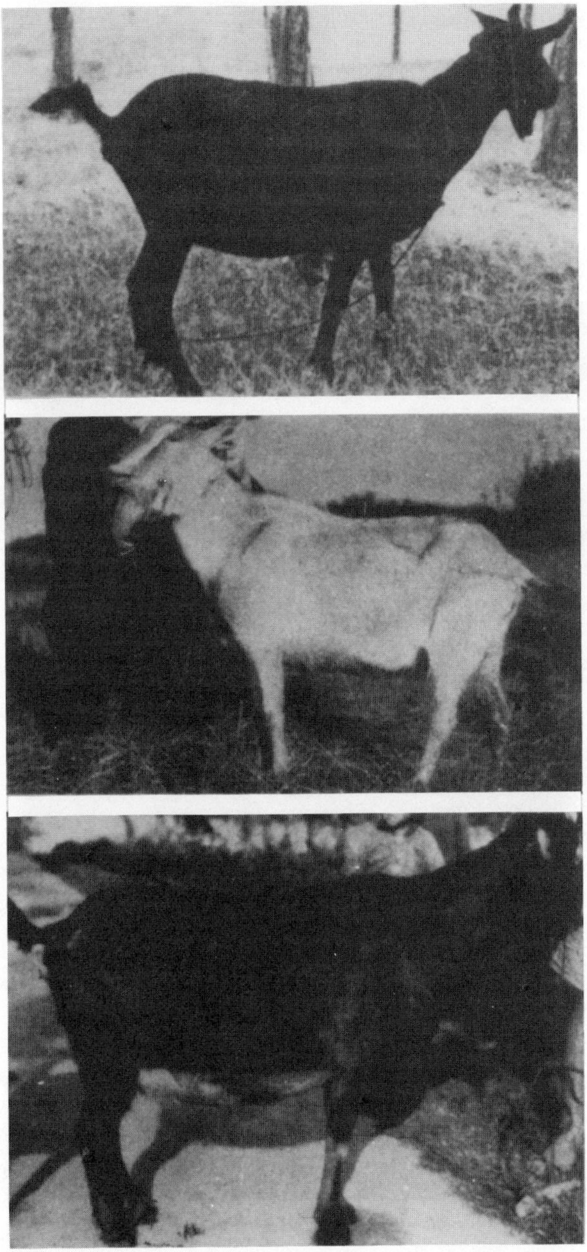

FIGURE 60.—Chinese goats are used primarily for meat, with skin and hair (from long-haired types) as by-products. Some types give a little milk. *Top:* Black doe of Szechwan. *Middle:* Fawn-colored doe of Szechwan. *Bottom:* Dark-brown "Mah" goat of the Chengtu plain in Szechwan.

and hair is short, long, and many variations in between. This is true in many other areas. Some of the small, long-haired goats observed

Figure 61.—*Top:* Black goats with medium-long hair. A type found in Shensi, Shansi, and possibly in adjacent areas. *Middle:* Short-, medium-, and long-haired goats of Chinghai. *Bottom:* Individual animals from a herd, part of which is shown in the middle picture.

in Chinghai are quite similar to the Kangra goat found in some areas on the southern slopes of the Himalayas in northern India.

According to Kislovsky (1938), the goats in Outer Mongolia range from 59 to 63 cm. in height at the withers and from 30 to 41 kilos in weight. They are described as a primitive type. He states that earless animals are occasionally seen.

Management and Feeding

The goat is kept primarily for meat. In some areas, especially in the northern portions of the country, they are used for milk production for family use. A limited number of commercial dairies and areas in which the goat is used by individuals for producing market milk have been developed. Before the present war a number of goat herds had been assembled around cities in North China. Most of these were in or near areas where purebred herds were available, and most of the animals in the commercial herds were grades from crossing purebreds and natives. Skins are an important by-product, and the hair is also collected from longer-haired animals.

The methods of managing goats are similar to those described for sheep. In the grassland areas goats and sheep frequently run together in a herd. In the farming areas the same situation exists. The goats are run in community flocks either separately or along with the sheep, or they may be taken care of by the individual owner, either by having some member of the family herd them during the day or by tethering them on grassy areas. Shelter may be provided in the owner's home or in adjacent quarters, or goats may simply be tethered near the house at night.

In southern Szechwan and certain other areas, some farmers allow their goats to graze only in the afternoon, owing to an unfounded belief that the dew on the grass will inflame the lips and cause diarrhea.

Few data are available on carcass yields in goats. Seven "Mah" goats slaughtered in Chengtu yielded the data, collected by Hsu (1943), which are summarized below. These included three wethers and four females, averaging 3¾ years in estimated age.

Item	*Average Catties*
Body weight	56.6
Carcass weight	25.4
	Percent
Carcass yield	44.6
Hide and hair yield	7.6

The feeds for goats are essentially the same as those for sheep. In the grasslands goats are on the range during the entire year, with no supplemental feed for the severe winter months. In farming areas they are allowed to graze along the hillsides, between fields, and any other places where grasses are available. During the winter months when grass is short, supplemental roughages, such as rice straw, other cereal straws in areas where rice is not grown, soybean straw, tree leaves, dried vines of sweet potatoes, dried pepper leaves and stems, etc., are fed. Grain is seldom fed, but in Yunnan it is reported that farmers give some grains such as rice bran, beans, and oats to their animals. The use of corn for fattening goats is also reported in this area.

In areas where goats are used for production of market milk, the feeding of grain is more commonly practiced. In and around Chengtu, for example, the materials commonly used for supplemental feeding are corn, soybeans, soybean dregs from making soybean milk, and bran. Bonemeal is also used to some extent as a mineral supplement.

Iodine deficiency has been found by Lu (1943) in Nubian goats at Chengtu. This deficiency resulted in a hairless or a nearly hairless condition at birth, stillbirth, or death very shortly after birth. Thyroid glands of the kids were much enlarged, and hyperplasia of the secretory cells was found. The addition of potassium iodide to the diet has corrected the deficiency, so that no more cases have been observed. Native animals in that area appear to be less susceptible to the deficiency, but this may be a cause of lowered reproductive efficiency in this and other areas where goiter is commonly seen.

Goats are slaughtered and sold for meat in the vicinities where they are produced. In the farming areas of China, goats are numerous and more popular for meat than are sheep. More are slaughtered during the winter than during the warmer months. The meat is sold in the usual small markets, where the carcass is hung up, cut into pieces as customers come to buy, and is usually retailed entirely on the same day it is slaughtered. Some goat meat is eaten with the skin on, after scraping off the hair.

Goat skins are an important by-product, and before the war many of these were exported. Shanghai was the chief point of export, with Tientsin second. Approximately 6,000,000 to 7,000,000 piculs were exported annually.[1] The skins for export were dried but were not cured in any way before being exported.

Some goat hair and wool are collected in certain areas, mainly in the northwestern provinces. This may be combed out in the spring or the animal may be clipped. The long hair is used for twine from which grain bags are woven, and the shorter, finer under coat is used for making felt. Some goat fiber was exported, and the extent of

[1] 1 picul=60.5 kilograms.

this, before the present war, is indicated in the summary which is given along with the discussion of the marketing of sheep's wool.

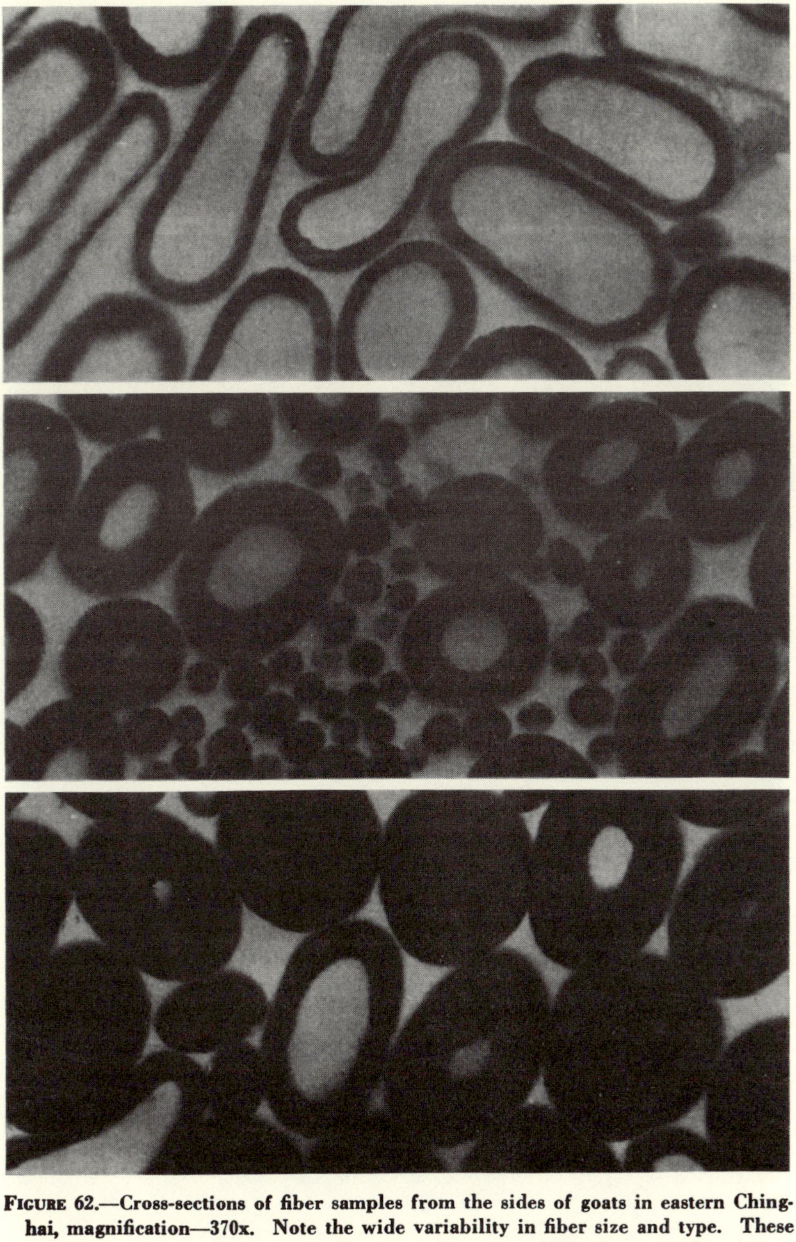

Figure 62.—Cross-sections of fiber samples from the sides of goats in eastern Ching-hai, magnification—370x. Note the wide variability in fiber size and type. These samples are from animals among the group shown in middle picture in figure 61.

Small intestines were collected in considerable quantities prior to the war and preserved for export for use as casings.

Milk is retailed in certain cities. With the exception of a few commercial dairies, this is largely individual enterprise. Methods of distribution are essentially the same as those described for cattle. Adulteration is just as prevalent as it is with cow's milk.

In one section of Yunnan (Lunan) a type of cheese is said to be made from goat's milk which resembles Roquefort in color and taste, except that it does not have green mold.

FIGURE 63.—*Top:* Goats at the livestock market held in connection with the regular market days (every three or four days) in Yungchang, Szechwan. *Bottom:* A Tibetan man uses a blower made of goatskin to start a yak-chip fire.

Breeding

No data are available on the age of puberty in native goats, but casual observations indicate that at least some females come in estrus for the first time at 4 to 6 months of age. Many farmers do not breed these young does until a few months later in order to allow better growth, but the usual practice is to allow breeding whenever estrus occurs.

Estrus apparently occurs at all seasons of the year in a part of the "Mah" goats which are found in the Chengtu area. Time of breeding was observed by Tang (1943) in 147 goats and the percentage of breedings in each season was as follows:

Autumn	*Winter*	*Spring*	*Summer*
41.5	12.9	23.8	21.8

Does would need to be maintained without breeding for an entire year in order to determine accurately the extent of the breeding season and the proportion that actually will show estrus at the various seasons. Lu (1943) reports that the most active breeding season is from about mid-September to mid-October, as judged by the large numbers of kids born in February, March, and April. Tang (1943) also obtained information on the length of the estrous cycle in 17 females. It varied from 13 to 29 days, and the average was 20.8 days. The distribution of the lengths of the cycle is given below:

Length of cycle (days)	13	18	19	20	21	22	25	29
Number of cases	1	3	2	2	4	2	2	1

Casual observations indicate that the estrous period lasts for from one to three days. The period of pregnancy is reported by Tang to vary from 139 to 152 days, and to average 148 days. Birth weights average from four to five pounds. Kids are weaned in the Chengtu area at 40 to 50 days of age.

The average number of young per birth is reported by Tang to be 2.09. In 56 births observed, the distribution of number of young per birth was as follows:

Number of young	1	2	3	4	5
Number of cases	12	31	10	2	1

The sex ratio of these young was 101.7 (58 females and 59 males).

Lu (1943) reports that approximately 85 percent successful breedings have resulted with the use of artificial insemination on goats in the Chengtu, Szechwan, area.

Data are not available on goats in other portions of China, so it is not known whether the data on "Mah" goats are representative of the conditions that would be found in other types.

In most areas castration is practiced, leaving only enough bucks intact to do the necessary breeding. In grassland areas the bucks run with the does the entire year. In farming areas some bucks are allowed to run with the does, while in other places the bucks may be kept separate. Bucks are kept only by one or a few individuals in some areas, and these bucks do the breeding for the other members of the community, for small service charges.

Little selection has been done with the objective of improving milk production or other characteristics. The buck is kept as a necessity to "freshen the females", but little thought is given to the selection of either superior bucks or does. Obviously, under these conditions, no definite breeding systems have been followed, and the native goats are generally inefficient producers of milk.

The introduction of improved breeds of dairy goats apparently was made by missionaries, who brought them to China to supply milk for their families and for use in grading up native animals. More development has taken place in northern and northwestern China than in the southern portions.

Some grading up has taken place around the cities and other places where herds of purebreds have been maintained, but very little actual data are available on the results obtained. Recent developments with goat breeding in Chengtu, Szechwan, indicate that the farmers are glad to have improved stock available for breeding. Madame Chiang Kai-shek arranged for the shipment of a few carefully selected Nubians from Hong Kong in 1939, and these have been maintained at West China Union University under the supervision of Dickinson (1943), where bucks have been made available for public service, and through the use of artificial insemination carried on by Lu and Lu (1943) large numbers of the local does have been bred. The offspring are larger and produce considerably more milk than the native goats, but actual data on production are not available, so definite comparisons cannot be made. They appear to be vigorous and thrive reasonably well under the local conditions but appear to be more susceptible to parasites than native goats, and this susceptibility may be a limiting factor in milk-goat improvement until preventive measures are available.

Saanens, Toggenburgs, Swiss, and Nubians have been imported, but no definite figures are available on the proportions of each breed.

Swine

Types

Chinese swine may be divided roughly into two main classes, the North China and South China types. Little is known of the history of these two types but it seems probable that they originated from somewhat different wild strains. Allen (1940) states that the number of recognizable races of wild pig in China and Mongolia can be determined only by further study of a large series of specimens. He reviews the available information and gives a tentative classification as follows: *Sus scrofa raddeana* in portions of Manchuria and Outer Mongolia; *Sus scrofa moupinensis* in Hopei, Shantung, Shansi, Shensi, Honan, southern Kansu, and much of Szechwan; and *Sus scrofa chirodonta* in Hupeh, Anhwei, and the other portions of China lying

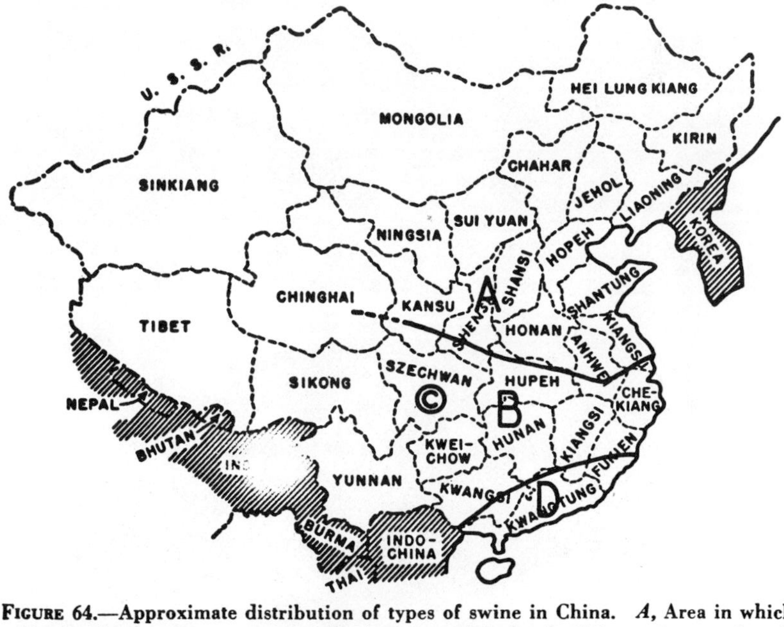

FIGURE 64.—Approximate distribution of types of swine in China. *A,* Area in which North China swine are found. *B,* Area in which South China swine are found. *C,* Area in which South China Yungchang or "bi-spectacled" white pigs are found. *D,* Approximate area in which Cantonese pigs are found.

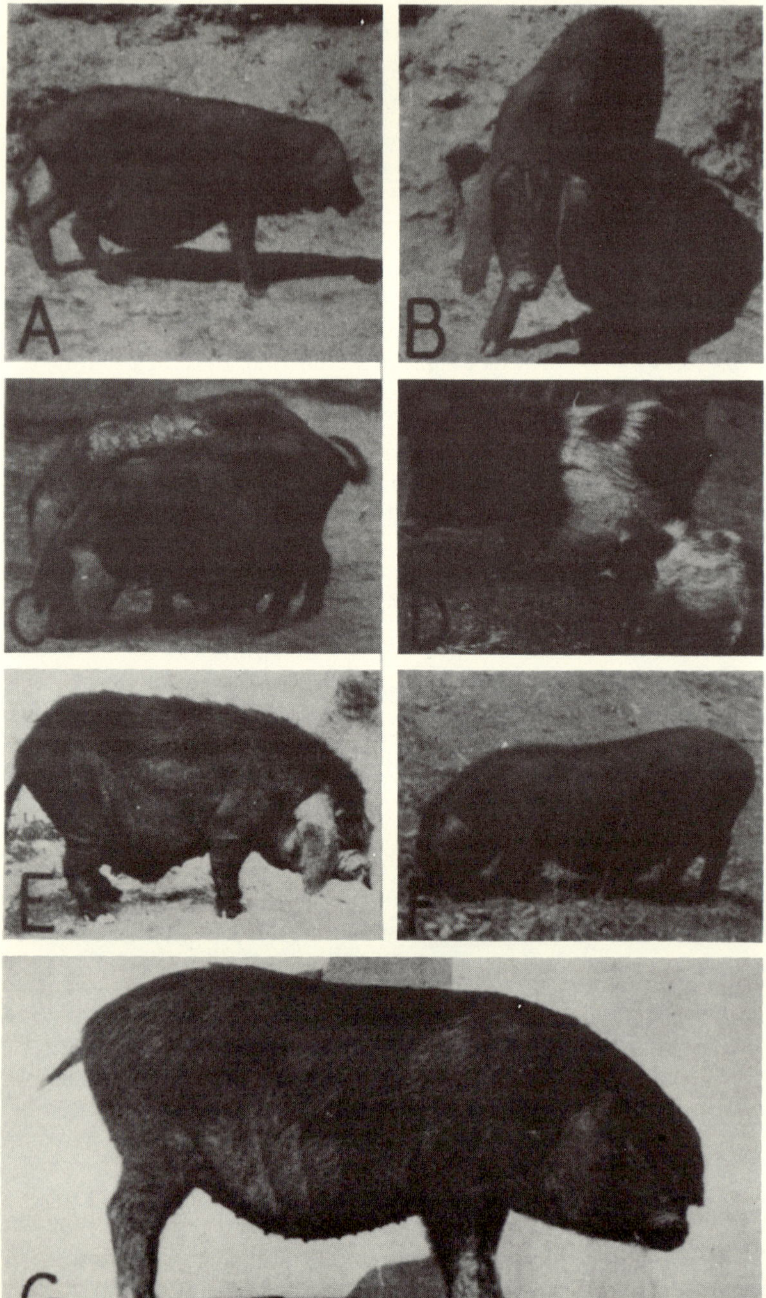
A
B
C
E
G

south and east of these provinces. The rough dividing line between the two latter types is approximately the same as the boundary between the North and South China types, shown in figure 64. However, Allen points out that these two wild types differ only slightly. There is not sufficient information available to judge whether the existing domesticated types are directly descended from wild types indigenous to the same areas.

The North China hogs are long, lean animals with long narrow heads, rather straight faces, and large ears. Black is the predominating color and white markings are rarely seen. The back is narrow and is usually straight or slightly sagging. The hams are very light. Wrinkles on the face are usually parallel and run along the long axis of the face. The rate of growth is relatively slow, but sows are reported to be quite prolific.

The South China pigs are shorter, broader-backed, more rotund, and the hams are somewhat larger, although not as plump as those found on most American and English breeds. The back usually sags more than in North China pigs. The head is short and rather wide, and the face is somewhat dished. Wrinkles on the face may be tortuous and tend to run at right angles to the long axis of the face. There is considerable variation in color in various areas, ranging from solid black to nearly white, and some animals are red or reddish-brown. The sows are reported to be less prolific than those of the North China types and the pigs grow more rapidly, but there is considerable variation among the South China pigs. The hair is said to be somewhat softer than on the North China pigs.

The approximate dividing line between the two types is shown on the map in figure 64. There is some intermingling of the types along this tentative boundary, of course.

The hogs of North China may be divided into two classes, a large and a small type. There are gradations between these two extremes, and the descriptive terms, large, medium, and small, are sometimes used in describing the variations in type.

Black strains predominate in many regions of South China, but

---

FIGURE 65.—North China swine. *A*, Sow photographed in Chinghai. This is the type commonly seen in Chinghai and Kansu. *B*, Front view of sow in *A*. *C*, Litter-mate sow and boar of Kansu, age 7 months. Sow in foreground was 11½ inches high at the top of the shoulder and was due to farrow in about three weeks. *D*, Small type of pig found in limited numbers in Kansu. (Photographed by F. P. Sha, Chinese Ministry of Agriculture.) *E*, Boar of the larger type found in Kansu and adjoining northern provinces (photographed by F. P. Sha). *F*, Pig of the type found in the vicinity of Chiangpei, Kiangsu. *G*, Sow of Shantung Province (photograph furnished by F. J. Rossiter, Office of Foreign Agricultural Relations, U. S. Department of Agriculture).

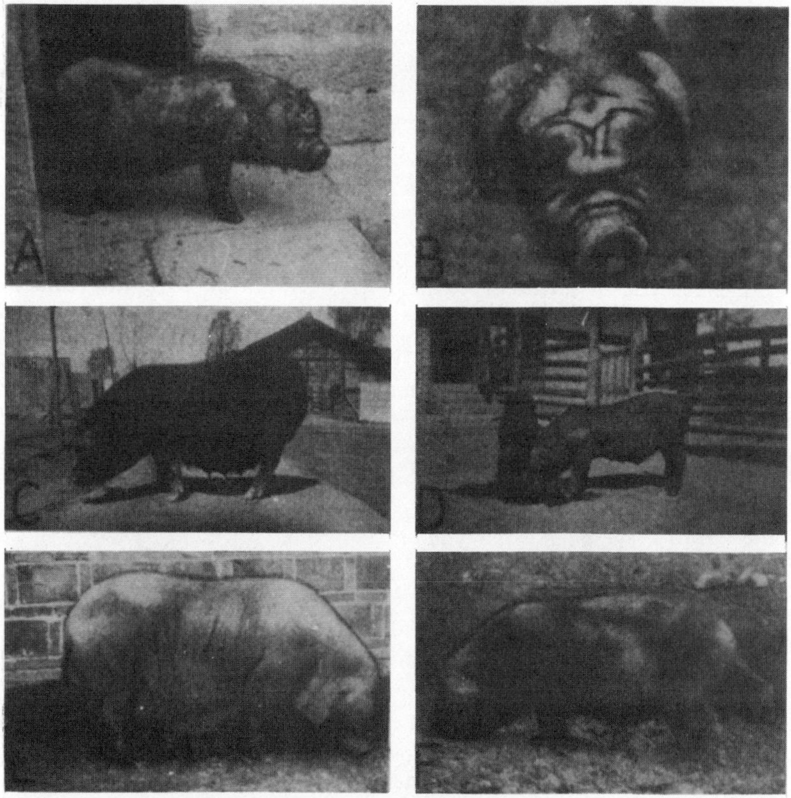

FIGURE 66.—South China black hogs. *A*, Small-type black boar of Szechwan Province. *B*, View of head of boar in *A*. *C*, Large-type sow of the type found in the vicinity of Neichiang, Szechwan, age 14 months (photographed by S. Y. Chang, Szechwan National University). *D*, Pigs of the same type as the sow in *C*. *E*, Sow of the type found in Kweichow. Red and brown animals are interspersed with the black ones in this area. *F*, Fattening hog of the same type as the sow in *E*.

white and black strains, such as Yungchang (in Szechwan) and Cantonese (in Kwangtung), predominate in certain localities.

Yungchang hogs are found principally in Yungchang hsien (county) and parts of Lungchang and Luchow hsiens to the south and west of Yungchang. (See figure 67.) These hogs are white over most of the body and have black spots around the eyes, varying from narrow circles to large spots that include much of the head and ears. They are sometimes referred to as "bi-spectacled whites". Some entirely white or nearly all white animals are seen. Adjacent to the home of the Yungchang hog there is a region where many black and white hogs are found, owing to crossing of the Yungchang with black strains.

Cantonese hogs are found in the vicinity of Canton and adjacent regions. There are not sufficient data available to outline this region accurately but the approximate area is shown in figure 64. The predominating color is black and white. The hogs in this area have been described by Levine (1920) who states that colors range from nearly white to black, or black with white points. In some localities the ears are large and pendulant, in others they are small and erect, and in some communities both types are found. It seems probable then that this is the area from which most animals were exported from China for use in swine-breeding in Great Britain to help refine the coarser British hogs.

Black hogs predominate in most other regions of South China, although some black and white pigs are found in Chekiang, Kwangsi, and Hunan Provinces, and animals with white markings are seen occasionally in other areas. Many of those found in Chekiang have a tendency toward "white belting", being white with black over the

Figure 67.—South China Yungchang "bi-spectacled" white hogs. *A*, Boar, age 2 years. *B*, Sow and litter. *C*, Sow, age 2 years, 10 months (photographed by S. Y. Chang). *D*, Weanling or "tethering" pigs. *E*, Fattening hog, age 6 months, estimated weight 75 pounds. *F*, Fat hog. *G*, Fat hogs nearly ready for slaughter. Age unknown, but had been in fattening pen 4 months and weighed an estimated 325 to 350 pounds.

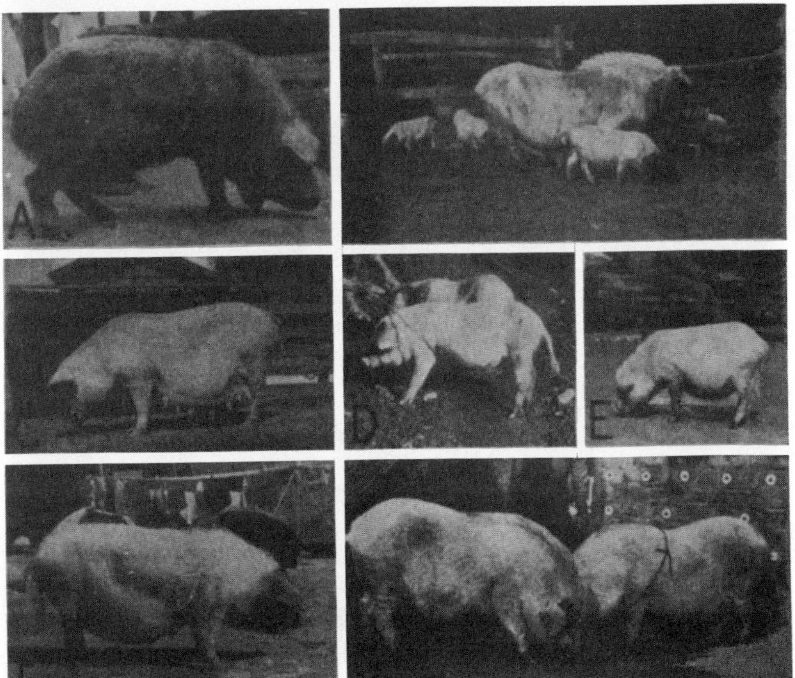

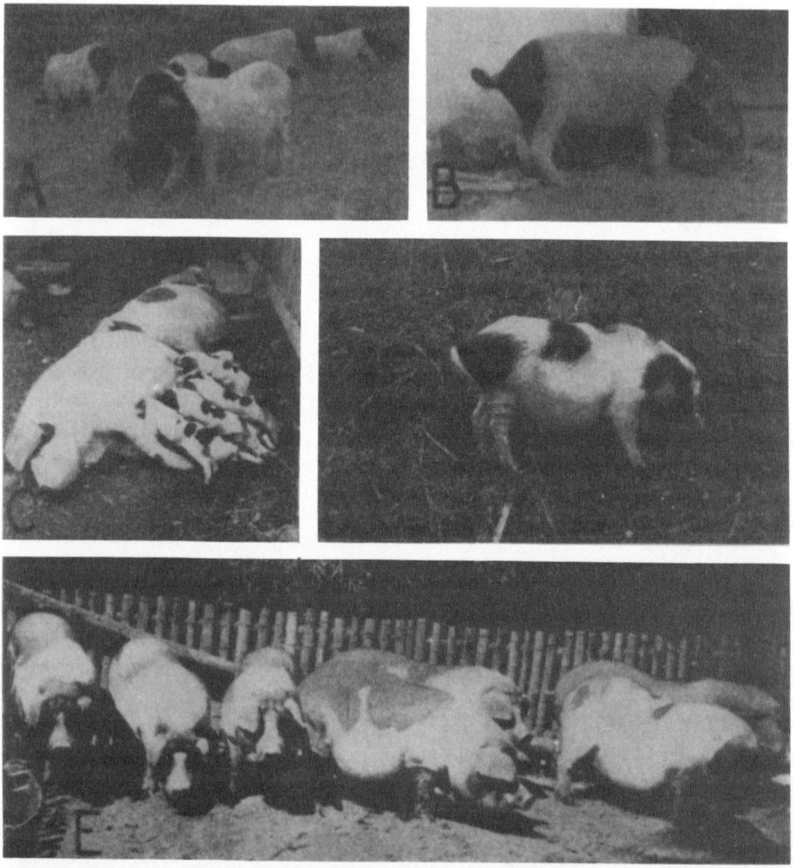

FIGURE 68.—South China black-and-white hogs. *A* and *B*, Pigs of a type found in vicinity of Chinhua, Chekiang. *C*, Sow and litter of a type found in vicinity of Nancheng, Kiangsi (photographed by F. J. Rossiter, Office of Foreign Agricultural Relations, U. S. Department of Agriculture). *D*, Spotted pig of a type found in limited numbers among the black pigs of Kweichow. *E*, Cantonese fat hogs (after Levine, 1920).

head and rear quarters. The natives refer to them as "two-ends-black pigs". Red or reddish-brown animals are frequently seen among the black strains, especially in the mountainous regions of Kweichow, Szechwan, and Sikong and similar regions to the north and south of this area. There is considerable variation in type within the black strains, even in adjacent areas. For example, the hogs around Neichiang, Szechwan, are larger, less chuffy, and somewhat later in maturing than those around Chengtu in the same province. Hogs in the mountainous regions tend to be somewhat leaner in appearance than those of the more fertile lowland areas, but this may reflect differences in environment rather than in heredity.

Pork is a favored article in the diet of the Chinese, and hogs are raised extensively in most areas occupied by Chinese. Chinese emigrants have taken hogs into Manchuria, mostly from Shantung, and these are described in the Japan-Manchukuo Yearbooks (see Wang, 1942) as being black and of three types, large, medium, and small. These appear to be of the North China type already described.

Many of the areas of Mongolia, Tibet, much of Chinghai and Sikong, and parts of Sinkiang have little feed that is suitable for hogs, hence the swine population is low. In Sikong the pigs are small and for the most part exist on vegetation. They are much smaller and have shorter ears than those of the Chengtu area, east of Sikong, in Szechwan. Chengtu is the chief city through which trade passes to and from Sikong. Mohammedans of the northwestern areas do not eat pork, so few hogs are raised in areas where this group predominates.

Owing to limited or total lack of available information concerning the swine in many localities of China, any description of types is necessarily incomplete. However, the foregoing descriptions and the photographs in figures 65 to 69 will indicate main types and some of the variations within these types. Other variations within type and details of subtypes can be filled in only as more information becomes available.

Management and Feeding

The place of swine in various types of farming and related activities in many provinces, such as Szechwan which has more hogs than any other province, may be summarized briefly as follows. Sows are kept by the relatively well-to-do farmers. The pigs are weaned at from 1 to 2 months of age and the majority of them are sold to poor farmers who feed them scantily for 2 to 5 months, then resell them as medium-sized feeders. These medium-sized feeders are fed mostly on roughage and farm refuse, to be converted into manure, after which they are again sold as large feeders. The medium-sized feeders are handled by small tenant farmers and others who desire them as a means of providing manure for their land. They sell these as large feeders to the well-to-do farmers and to proprietors of mills, oil presses, breweries, etc., to be fed on fattening rations for two or three months, after which they go to market as slaughter hogs. The total feeding period, from weaning to slaughter, may vary from six months to as much as two years. There are many variations of the above pattern, of course, but it serves to illustrate the essential pattern of swine production, in which animals are "stretched up" largely on roughage and under otherwise poor conditions of environment, after which they

615999—45——9

FIGURE 69.—Hogs of a type found in limited numbers in Sikong. They exist largely on grass and other roughage.

are fattened for a brief period on relatively good-quality feeds. Available experimental evidence (McMeekan, 1938) would indicate that this procedure tends to produce animals with relatively underdeveloped muscles and a high proportion of fat.

The practices described above are not universal, however. In Shansi, for example, the common practice is for a farmer to buy a weanling pig and feed it to maturity or slaughter weight.

Pigs may be housed in one portion of the farmhouse or in a shed built adjacent to it, or in one portion of a mill or other commercial establishment. The younger pigs may be tethered or allowed to run free during the day and put in the house or pen at night. The larger feeder pigs and the fattening pigs are usually kept confined. The pen in which they are kept may have a pit underneath the floor in which the feces and urine accumulate. This type is used where wood is plentiful. Others have a sloping stone floor with a pit at the side or outside. In still others the pigs may be put in a pit to which bedding is gradually added to keep it relatively clean and to soak up the urine. In the less intensive farming areas the pigs may be turned out to graze during the day and be put in their pens at night, or they may be allowed to graze day and night. In the more hilly regions, feeder hogs are herded during the day, being allowed to graze in harvested fields, on grassland, along streams, and in wooded areas.

When put on a fattening ration, hogs are usually confined in a pen and remain there until they reach slaughter weight. The fattening pens are usually built in the back yard or in any secluded corner on the farms. On ordinary farms, not more than one to three hogs are usually fattened at one time. Manufacturing plants may have a series of pens in one portion of the plant or in an adjacent building. Feeding in these plants is often on a seasonal basis. For example, the sugarcane plants operate for about four months during and after harvest and do most of their fattening of hogs during this period. Pigs are usually fattened in dark places without sunlight, and quarters are usually very unsanitary. Hsu, Peng, and Feng (1938) report work which indicates that large feeder hogs can be fattened successfully in such confinement, apparently because the feeding period is not sufficiently long to exhaust their store of vitamin D and because they are making less bone growth than younger pigs, but that the method of confinement without sunlight was not so satisfactory for the younger feeder pigs. In Chekiang, hogs are kept in a very dark room and the feed may be thrown in on top of the manure for the pigs to sort out.

In Szechwan most sows are bred to farrow two litters a year. Parturition may occur at any time during the year, but the main farrowing seasons are from February to May for the spring crops and September to December for the fall crops. Thus the pigs are born at temperate seasons, for the most part, avoiding both the hot and the cold seasons for the suckling period.

There is much variation in feeds and feeding methods. There are certain practices that prevail rather generally, however. There is a preponderance of coarse, high-fiber and low-protein content feeds in the diet of the growing pig. Fattening pigs receive a much more concentrated ration but the protein content is low. Grains are usually cracked or ground before feeding, and cooking of the feed is commonly practiced. Slop feeding is the rule. The calcium and salt contents of the ration are usually low.

Figure 70.—There are many types of structures used for hog houses in China. Dr. E. A. Tunnicliff, of Montana State College, inspects one of the more elaborate type used in Kweichow. Note the pit underneath, in which feces and urine accumulate.

Feeds that are in common use are listed below:

Grains and seeds
- Barley
- Broad bean
- Buckwheat
- Buckwheat flowers
- Corn
- Kaoliang
- Peas (field)
- Rice
- Soybeans

By-products
- Brewers' grains
- Garbage, solid
- Garbage, liquid (mainly rice water)
- Peanut cake
- Rapeseed cake
- Rice milling by-products
- Sesame cake
- Soybean cake
- Soybean curd ration and residue

By-products—Continued
- Sugarcane scum or residue
- Sweet-potato flour residue
- Wheat bran

Tubers
- Potato
- Sweet potato
- Taro

Roughages
- Broad-bean chaff
- Herbages
- Soilages
- Sweet-potato vines
- Taro tops
- Vegetable refuse
- Vetch-leaf meal

Other
- Infertile and dead chick eggs from hatcheries

The practice of cooking feeds has been shown by Hsu, Peng, and Feng (1938) to be undesirable. In two experiments in which cooked and raw feeds were compared, pigs grew just as rapidly and consumed 11 percent less feed per 100 catties of gain on the raw feed, not considering the labor and cost of fuel for cooking the feed. They conclude that the cooking of naturally dry feedstuffs should be discouraged. This is in accord with results in other countries. Other feeding trials conducted by Hsu have brought out the following points:

Rapeseed-oil meal was proven by Hsu (1941) not to be a dangerous feed for swine. It could be used up to 30 percent of the ration without ill effects. Ten percent was advised for all classes of swine to remedy the protein deficiency of the typical Chinese swine rations. When supplemented with 8 percent rapeseed-oil meal and 1 percent salt and bone meal, the typical farm ration gave quicker and more economical growth than when fed without these supplements.

Rice bran alone was shown by Hsu (1941) to be entirely unsatisfactory as a feed for foreign purebreds. The maximum that could be fed to these animals was 25 percent of the ration, but for native hogs up to 50 percent could be used. Peanut-oil meal was shown to be a much more satisfactory protein supplement than rapeseed-oil meal. Sugarcane scum and yellow corn, commonly used by farmers in certain areas, proved to be more economical than corn and rice bran which is also commonly fed. Supplementing this ration with peanut-oil meal resulted in even faster and more economical gains. Supplementing with similar amounts of rapeseed-oil meal did not give as satisfactory results as the peanut-oil meal.

Cracked rice and rice polish are used extensively in the Yungchang area for fattening hogs. Hogs made slower but more economical gains on a rice diet than when fed without a supplement of yellow corn (Hsu, 1941). Rice and rice polish fed with distillers' grains, as is practiced by some farmers in this area, proved to be an ineffective and expensive type of ration.

Feeders that are bought for fattening are usually placed in pens with four to six hogs in each group and given a fairly coarse ration at first. They are gradually shifted over to the fattening ration. In the sugarcane district the fattening period seldom exceeds three months and averages about two and one-third months. During this period animals are given all they will clean up twice daily. After a long growing period without adequate feed, they put on gains at a rapid rate, gaining from 30 to 50 kilos during the feeding period.

Sows are usually fed, except when nursing, on a very rough ration. It includes such items as broad-bean chaff (stems and leaves), vetch meal, fresh vegetable refuse, rice or wheat bran, brewers' grains, and kitchen refuse. Feeds for sows are not usually cooked. Sows become quite thin and emaciated on the rations usually fed.

Since Szechwan is the largest swine-producing province, a review of marketing procedures in that province is given to indicate market procedures. Somewhat different procedures are used in some other areas.

A pig usually changes hands several times before it reaches the market as a fat hog ready for slaughter. First it is sold as a weanling, then as a small feeder, followed by later feeding stages, and finally as a butcher hog. Most of these transactions take place at market gatherings held at regular intervals in the city or town that serves as the marketing center for an area. At these market places, there is a hog yard where sellers, prospective buyers, and middlemen gather. Marketing usually begins in the early morning. Weanling pigs are bought in basketfuls, tetherers and small feeders in pairs, and larger feeder hogs in singles or bunches. Butcher hogs do not come to such markets. They are sold in the pen direct to butchers or professional buyers, except in a few large cities, such as Chungking and Chengtu, where special fat-hog markets are provided.

Bargains are made in various manners. In most places hogs are bought and sold on the head basis. The owner asks so much for his animal and the buyer makes an offer, both to the middleman with finger signals in the sleeves. The middleman or commission man then goes through many motions with much loud talking until a deal is closed. He usually gets 2 to 3 percent, borne either by both parties or by one, depending on local custom. Since neither party knows what goes on in the sleeves between the middleman and the other

FIGURE 71.—Weanling and feeder pigs of various ages are sold on regular market days. *Top:* Weanling pigs in baskets at the market in Yungchang, Szechwan. *Bottom:* Buyer and seller bargain at the same market. The two men, apparently shaking hands, are bargaining in their sleeves, where prices are indicated by finger signals.

party to the deal it is easy for the middleman to cheat both, getting more than his stipulated commission. Since ownership often changes so many times during the life of a pig, these unregulated middlemen place quite a financial burden on swine producers.

In some localities pigs are sold on a weight basis, in which case the bargain is on a per catty basis. In some such cases the actual weight is not determined but is merely estimated.

Most weanling pigs are taken immediately after weaning to the market for sale. A few are retained for breeding and some are sold directly to neighboring farmers. The smaller pigs command higher prices per unit weight than larger ones. Weanling pigs usually range from 10 to 20 catties per head. Weanling pigs are usually tied with a straw rope around the neck or chest for convenience in handling at the

FIGURE 72.—Fat hogs are brought to market in many ways, including driving, carrying on a coolie's back, carrying on two poles after the fashion of a sedan chair and on wheelbarrows. Animals are slaughtered after midnight, pork is on the market in the early morning and must be sold that day since there is no refrigeration. *Top:* A fat hog on a wheelbarrow in one of the fat-hog markets in Chengtu, Szechwan. *Bottom:* Meat market in Chengtu. Pork is the chief type of meat offered in meat markets in areas where Chinese predominate. There is also some goat meat and beef.

market and afterward. They are fed reasonably well for the period following weaning and may be taken out on leash by children or old women to graze. But the level of feeding is not sufficient to maintain good growth, and after a month or so, even though they have gained a few catties, they have changed from plump, slick-haired weanlings to pot-bellied, rough-haired animals with relatively large heads, bulging eyes, and weak limbs. Some are again led to market for sale as tethering pigs. Whether or not they are resold at this stage, they are then allowed to roam about, picking up anything that is edible; they may be given some liquid garbage, and sometimes a little rice bran may be added to the ration. They are then ready for resale as feeder pigs.

Most feeder pigs are sold at market gatherings. They are rarely weighed. They are divided into three categories, small, medium, and large. These have the following approximate ranges in weight: 40 to 70; 70 to 100; and 100 catties up, respectively.

In rural areas buyers purchase the fat hogs direct from the farmer while they are still in the feeding pen. The buyer estimates the live weight and bargains on this basis. Actual weights are seldom taken. Since the buyer is trained in evaluating weights and carcass yields, the farmer is at an obvious disadvantage in these deals.

The large consuming centers maintain fat-hog markets. For example, there are four in Chengtu. These have a large central open space, with open sheds built around the four sides. Each yard has an owner or manager in charge, one clerk, and a number of commission men. Markets are held at regular intervals, and all do not operate on the same day.

Fat hogs are taken to market in several ways. One or two animals may be strapped on their backs on a wheelbarrow, with leaves or straw under them for padding, and pushed to market. Boats are used along the streams that lead to markets, 20 to 30 hogs being jammed in one boat. Animals are also carried, either suspended on a board framework or on boards on top of two poles after the fashion of a sedan chair, being taken to market by two men. Animals are also driven to market if the distance does not exceed five kilometers. One or two or more up to the point of good-sized droves are taken in this way. Heavy hogs walking to market often become footsore and are shod with straw shoes. The ears may be fastened back to the head by tying or sewing together to enable the animals to see better, and in some cases the eyelids are also suspended by sewing. Another method is for a man to carry a hog, strapped to his back.

Prices vary with the type of animal and the season. Small weanling pigs command the highest prices and large feeders the lowest, per unit weight. Top prices are received at two seasons, March and April, and November and December. These are the busiest farrowing periods, hence there is the smallest number of weanling pigs for sale. Prices of large feeders and pork do not vary as much within the year as do prices of the smaller animals.

Gilts for breeding are usually bought while still in the litter, or at weaning time. Sows seldom change ownership. Occasionally an old, run-down sow or a pregnant one is brought to market gatherings. Pregnant sows bring a much higher price than non-pregnant ones.

There is very little market for boars. Young boars are discarded from breeding, castrated and fattened, before being placed on the market.

Methods of slaughter will not be discussed in detail here. There are many local variations, of course, that would require careful observation to describe in detail. In some places, the practice has developed of pumping water into the carcass to increase the weight. This bad practice is in the same degraded category as putting water in milk to increase the volume. Slaughtering usually begins about midnight, so the carcass can be in the retailer's hands early the next morning. The carcass is exposed in open-air markets until sold. It is cut in strips, diagonally across the carcass, and practically all parts of the carcass command the same price. The head and intestines sell for about one half the price of the main portion of the carcass. Fat usually brings up to twice the price of the carcass itself per catty. Kidney fat, for example, usually brings 1.5 to 2 times the price of fresh pork.

Most pork is sold in the fresh, unchilled state and with the skin on. Meat is also cooked with the skin on. Some hams are cured, and prior to the war some of these were exported. These hams are usually cured in the fall and winter. The native method is to salt the ham, using 1 pound of salt per 16 pounds of meat. It is then sunned for about one week and hung up in a cool place. Mold grows on the surface the next summer. It is usually eaten after about one year but is considered best after two or three years. The flavor is quite strong.

Bristles

Bristles are an important by-product of the slaughterhouse. Those along the withers and backbone are usually plucked and bundled separately by butchers. Longer bristles from along the sides of the backbone are scraped off black hogs, and some hairs from other parts that are not suitable for bristles may be scraped off and bundled with them. In white hogs, the bristles along the back and on either side are all plucked out. This accounts for the higher yield of marketable white bristles and for certain differences in the processing of the two colors of bristles. Generally, 80 to 90 catties of finished product are obtained from each 100 catties of white bristles, and 65 to 70 from each 100 black ones.

Bristles are collected from butchers by jobbers or by small shops in the villages and sold to city dealers in raw condition. They must be dried to avoid spoilage during transport and storage. Jobbers usually do some sorting on the basis of length to make the bundles more uniform and attractive to dealers. Final processing is usually done in large cities, of which Chungking is the most important center.

The processing of black bristles is relatively simple, the essential steps being as follows: soak in water for 10 to 20 days to loosen dirt and to allow attached particles of skin to degenerate; wash, steam, dry, comb out soft bristles, straighten on thin boards, dry, then grade into length classes and put in rolls.

The processing of white bristles is much more elaborate. They are soaked in water for 24 hours, then bleached with sodium peroxide. After this they are exposed to sulfur fumes, dried, and the black and yellow bristles picked out. This is followed by a second and a third bleaching, after which they are again exposed to sulfur fumes. They are then de-sulfurized with sodium hyposulfite, after which soft bristles are combed out, straightened, and placed on bamboo slats, and dried. They are then ready for grading into length groups, inspection, and packing. The Yungchang hog, described earlier, is perhaps the most important source of white bristles.

There are, of course, some variations of the processes outlined above.

Normal bristle production in Szechwan before the present war was about 16,000 piculs (1 picul = 60.5 kilograms) of black and 3,000 piculs of white ones per year. Actually about 6,000 piculs were brought in from Yunnan and Kweichow, and 4,000 from Shensi and Kansu provinces. They were gathered along the two main water routes, the Yangtze and the Chialing Rivers.

Bristle exports have decreased during the war, and owing to wartime conditions no distinction is being made by foreign buyers between black and white bristles. Owing to the greater cost of processing the white ones, emphasis is now being shifted to the black bristles for current export trade.

In the dressing or grading of bristles into length groups, they are usually sorted into bundles ranging from 1½ to 6 inches, by ¼-inch gradations. Sometimes finer distinctions are made. A few bristles in excess of 6 inches are found but they are of minor importance.

Prior to the war the carcass of a slaughter hog yielded the butcher about $30 and the bristles about $6 to $7. Thus the bristles accounted for approximately one sixth of the value of the animal.

Pig skins are used in some areas for the making of leather for suitcases.

Breeding

Boars are usually kept by poor or disabled farmers. In some areas the keeping of a boar is looked down upon as a lowly occupation. The boar travels in some localities, while in others the sow is brought to the boar for breeding. Each traveling boar covers an area about five kilometers in radius. During the rush season a boar may serve five to six sows per day. The owner usually is required to pay a service charge at the time of service, and after the sow is either pregnant or has given birth, an additional charge equivalent to the current price of one catty of weanling pig. Boarkeepers are sometimes referred to as boar chasers because they follow their boars from one farmhouse to another. Their usual practice is to buy a male pig from a farmer, and it is usually one of the smaller ones from a litter since the small ones cost less and will eat less. The boar is raised on the same poor diet as is common for tethering and small feeder pigs, consisting largely of roughage, until about 5 months of age. After that it serves regularly until 10 months or a year of age and then is castrated and fattened for market. Young ones are easier to drive and to maintain and are more apt to be willing breeders.

Farmers usually feed their developing breeding gilts very poorly, yet they seem to survive and give reasonably satisfactory breeding performance in spite of this poor treatment. An experiment conducted by Hsu and Peng (1942) in which gilts were kept on a well-balanced ration, as compared with the typical farm ration, indicated that the well-fed gilts would produce more and heavier pigs per litter at the first farrowing than the poorly fed gilts. Differences were less marked at the second farrowing. Individual sows are usually kept in pens of about 10 x 10 feet and the floor is either dirt or is paved with stone or board sloping to one side for drainage into a gutter from which refuse collects into a tank. Sows are confined to this pen except when removed for mating or to be sold.

Farmers seldom bother to remember breeding dates or to calculate the expected time of parturition. Some straw bedding may be thrown in at the time the sow appears to be ready to farrow, but otherwise the sow is left largely to herself during and for several days after farrowing. Many farmers do not even know the size of the litter or the sex of the pigs until well after farrowing. Luckily, the Chinese sow appears to be a good mother, and death losses during the suckling period are relatively small.

Spaying and castration are both commonly practiced. This is usually done at 4 to 6 weeks, by local practitioners who charge so much per litter for performing the operation. Castration may be done on the farm, while spaying is frequently done at the market at the time the weanling pigs are being sold, or both operations may be done at once on the farm. The practitioners are very clever at the spaying operation, accomplishing it very quickly.

The work reviewed below was conducted at National Central University and was largely financed by a grant from the Rockefeller Foundation in connection with its rural-reconstruction program in China. From February 1936 to October 1937 the work was conducted at Nanking. Owing to the Japanese invasion, the staff and part of the herd were moved to Chungking and then to Chengtu. After the move to Chengtu, the Szechwan Provincial Government cooperated with National Central University in the work until it was concluded in 1941.

Owing to wartime conditions some of the data were lost; the work is inadequate in some respects, and not as complete as could be desired in other respects, but these appear to be the only data of their kind in existence. Much of the information has been published in Chinese by Hsu and co-workers.

The Chinese swine used at Nanking were of the North China type, with the exception of a few from Chekiang that were intermediate, while those at Chengtu were of the South China type. Rations used at Nanking were comparable to those commonly fed under conditions of good management in the United States, but tankage was not available at Chengtu so only vegetable protein was supplied.

A survey of levels of fertility on farms in four sections of Szechwan, including the Chengtu, Yungchang, and Neichiang areas, revealed that litter size varied from 8.4 to 9.1 in the four areas and averaged 8.8. In 31 sows native to the Chiangpei area of Kiangsu, an average litter size of 13.9 was observed. Tong (1943) cites an exceptional case of a North China black sow, kept at Minghsien College, that had 32 tests and produced 25 live pigs in one litter. Levine (1920) states that in the Canton area a sow is supposed to be able to nurse at least 10 pigs, which seems to be the average for the region, although as many as 15 are frequently seen with one sow. Thus it appears that the Cantonese sows are more prolific than those in some other parts of South China.

Chinese pigs are reputed to develop sexually more rapidly than modern breeds. Chinese animal-husbandry workers cite instances of gilts being bred by litter-mate boars before 60 days of age. Gilts are usually bred when about 5 months old, but earlier breedings are common, and Chinese workers say that it is not unusual to find gilts farrowing at about 6 months of age. This is in marked contrast to the findings of Phillips and Zeller (1943) with the Poland China breed, in which spermatozoa were first observed in the testes at 20 weeks of age, and in which large- and small-type gilts exhibited first estrus at 198.7 and 207.8 days, respectively. Hsu and Peng (1942) studied 10 gilts of the type found in the vicinity of Neichiang, Szechwan. Five were on a standard and five on a typical farm ration and the average ages of first estrus were 88.6 and 108 days, respectively. First estrus averaged 5.2 and 5.8 days in length and later estrous periods were shorter, the average for the groups varying between 2.0 and 3.0 days after the second estrus. These gilts were checked for estrus twice daily.

Pigs from Chiangpei in Kiangsu Province and Chinhua in Chekiang were compared and both types were about equal in rate of growth. The Chinhua pigs had more subcutaneous and less kidney fat than those from Chiangpei. These regions are on the borderline between the North and South China types. The Chinhua pigs, owing to the thicker back fat, appear to be more like the South China type than those of Chiangpei, and the black and white markings indicate that they may be intermediate between the North China and a black-and-white strain of the South China type.

Relatively few data are available concerning the feed-lot performance of the native pigs. Some of these data will be summarized along with the data on crosses with American and English breeds. Some data, not involved in those studies, are summarized here.

Four pens of pigs purchased in the Chiangpei, Kiangsu, area were fed for varying periods, and the results are summarized in table 2.

A direct comparison was not made with pigs from the vicinity of Chinhua, Chekiang, but six of the latter type were fed in an experiment to compare growth rates of pigs confined in sunlight and in a dark room. This work was stimulated by the common practice among Chinese farmers of confining their pigs in dark quarters. The data are summarized in table 3.

A comparison of the performance of three strains of Szechwan pigs by Hsu and Chiang (1940) showed that the types differed in time of fattening and in weight when it was most profitable to market them. Small, black pigs from Chengtu were at their optimum conditions at 180 catties[1] (about 198 pounds), larger black pigs from Neichiang at 250 catties (about 273 pounds), and Yungchang pigs at 210 catties (about 231 pounds). At lower weights they were not fat enough to please the public demand for a fat carcass, and if fed longer the gains became too expensive. Some data obtained on animals being slaughtered in the normal course of business in the Chengtu, Neichiang, and Yungchang markets are summarized in table 4.

Hsü and Peng (1941) studied the performance of six sows from the Neichiang, Szechwan, and eight from the Yungchang, Szechwan, area, including performance of the sows and their progeny. The data

[1] This is the new catty, equal to ½ kilo or about 1.1 pounds. The old catty was equal to about 1.3 pounds.

are summarized in table 5. The pigs were weaned at 10 weeks and put on feed at 13 weeks of age. Average or medium-sized pigs in each litter were selected for feeding, the extra large or small ones being discarded. The number of pigs farrowed is much higher than that cited earlier as the average for farm sows of this type. These sows were especially selected and were maintained under much better conditions than are commonly found on farms. The Neichiang pigs grew at a more rapid rate over the entire feeding period but the greatest difference occurred after the pigs reached 220 catties. The Neichiang pigs made more efficient gains, and the difference of 67.6 catties in feed required per 100 catties gain was statistically significant. The carcasses of the Yungchang pigs were somewhat fatter, owing to slower growth and a longer fattening period.

TABLE 2

PERFORMANCE OF FOUR PENS (FOUR PIGS PER PEN) OF NORTH CHINA (CHIANGPEI) PIGS

Items Compared	Pen Number			
	1	2	3	4
Days on feed	37	74	92	126. 0
Initial weight [a]	87. 8	85. 2	84. 5	88. 8
Final weight	120. 5	152. 0	179. 6	179. 4
Average daily gain	0. 88	0. 90	1. 03	0. 69
Feed per 100 catties gain	444. 7	455. 0	470. 2	592. 2
Dressing percent (warm)	68. 8	70. 2	74. 1	75. 4
Back fat, cm [b]	3. 05	4. 38	5. 0	5. 12

[a] In catties.
[b] Measured at top of shoulder.

TABLE 3

PERFORMANCE OF CHEKIANG PIGS IN SUNLIGHT AND IN DARK QUARTERS

(Three pigs in each lot)

Items Compared	Conditions	
	Sunlight	Darkness [b]
Initial weight [a]	38. 2	36. 7
Days on feed	133	133
Final weight	149. 7	139. 2
Average daily gain	0. 84	0. 7
Feed per 100 catties gain	400. 6	442. 6
Dressing percent	72. 3	71. 9
Back fat, cm.	5. 60	5. 25

[a] In catties.
[b] Only 2 pigs finished this trial.

TABLE 4

SUMMARY[*] OF SLAUGHTER DATA IN THREE SZECHWAN MARKETS

Items Compared	Chengtu			Neichiang		Yungchang	
	March	April	May	Winter	Spring	Winter	Spring
Number of hogs	19	35	30	37	35	28	16
Live weight [a]	169. 7	181. 5	173. 2	237. 8	272. 4	221. 0	196. 2
Dressing percent	72. 2	66. 1	68. 2	78. 7	79. 0	75. 4	72. 4
Caul fat (percent)	1. 79	1. 71	1. 92	4. 80	5. 20	4. 73	4. 09
Kidney fat (percent)	2. 84	2. 78	3. 05	5. 02	5. 27	5. 33	4. 23
Back fat (cm.)	5. 73	5. 67	5. 55	6. 87	7. 51	8. 13	7. 65
Length of bristles (cm.) [b]	9. 11	8. 81	8. 04	12. 39	11. 29	13. 44	11. 79

[a] In catties. Catty = ½ kilo, or about 1.1 pound.
[b] Bristles on back.

TABLE 5

AVERAGE PERFORMANCE OF SIX NEICHIANG AND 8 YUNGCHANG SOWS AND THEIR PROGENY

(Four pigs per litter were fed)

Items Compared	Breeding of Sows and Pigs	
	Neichiang	Yungchang
Pigs farrowed	12. 7	12. 0
Pigs weaned	10. 3	8. 5
Average birth weight (catties)	1. 5	1. 5
Average weaning weight	26. 0	24. 6
Average initial weight in feedlot	31. 0	27. 8
Average final weight	243. 7	240. 2
Average final age (days)	331. 3	387. 0
Average daily gain	0. 89	0. 72
Feed per 100 catties gain	506. 7	574. 3

Levine (1920) reports that 12 hogs fed at the Canton Christian College made an average daily gain, at 6 months of age, of about 0.65 pound a day on full feed, and that these pigs usually weighed between 200 and 250 pounds at 1 year of age. The average dressing percent of 32 hogs was 72.5.

The studies in which native pigs were compared with modern and modern × native crossbreds are summarized below. Owing to the different native types used in work at Nanking and Chengtu, the work at the two locations is reviewed separately.

Work at Nanking. The modern breeds involved in this work were Berkshire, Chester White, Duroc-Jersey, Hampshire, and Large Yorkshire. The Duroc-Jersey and Yorkshire hogs were rather large and rangy, while the hogs of the other four breeds were of a smaller, more compact type. Native pigs were of the North China type and came from the vicinity of Chiangpei, Kiangsu. Numbers were limited and the modern breeds were grouped together for some comparisons with the native pigs. Loss of original records, owing to wartime conditions, makes it impossible to indicate the proportions of the various modern breeds in some of these cases. All pigs were given comparable treatment, and the management and feeding were as near to generally accepted American standards as the working conditions permitted. The ration generally used included corn, tankage, soybean-oil meal, bonemeal, and salt. In feeding trials with fattening hogs, the protein content of the ration ranged from about 16.6 percent at the beginning to about 12.9 percent at the end.

The duration of gestation was similar in all groups. The average length of 32 gestations in modern sows was 115.8±0.51 days; in 21 gestations in native sows it was 115.1±0.39; and in five F_1 sows mated to modern boars, it averaged 113.6±0.87.

The interval between deliveries of individual pigs at parturition in the above modern and native sows averaged 23.2±4.18 and 19.7±2.69 minutes, respectively, and the difference was not statistically significant.

Pigs from sows of modern breeds were somewhat heavier than those from native sows at birth. The average weight of 334 pigs of modern breeds was 2.06±0.04 catties. Eighty-eight native pigs averaged 1.61±0.04 and 314 crossbred pigs out of native sows averaged 1.69±0.03. Differences between various groups are shown below, with their standard errors:

Modern–Native	0.45±0.057 catties
Modern–F_1	.36±0.049 "
F_1–Native	.08±0.049 "

The native sows were smaller and had larger litters than the modern sows, which at least partially accounts for their pigs being lighter.

The 462 male pigs born in this work averaged 1.86 catties and the 346 female pigs 1.84 catties at birth, and the difference was not significant. The sex ratio was 133.5.

The Chinese sows are reported to be good mothers and the available data bear out this belief. Data on survival to weaning (at 10 weeks of age) are summarized in table 6.

The changes in weight of the sows owing to reproduction indicates that the loss is greater in the native sows. Sixteen modern sows averaged 317 catties before parturition. They lost an average of 40.8 catties during parturition and 38.4 during the 10-weeks lactation, a total of 79.2. Eighteen native sows averaged 285.7 catties before parturition, lost 47.1 and 46.0 catties, respectively, at parturition and during lactation, or a total of 93.1.

The Chinese sows, being smaller, consumed less feed than the modern sows during lactation. During 1936 the average daily feed consumption of four sows of each type was 9.93 and 12.1 catties per head. Duing 1937 the sows had access to green alfalfa and the consumption of concentrates was somewhat lower, averaging 8.84 catties daily for 12 modern sows and 7.73 for 11 native sows. An interesting comparison was made of the combined feed consumption of the sow and pig during the nursing period. The data are summarized in table 7. The difference in efficiency of production, thus measured, was 94.9 ± 21.29 catties. The native sows were raising hybrid pigs, however; hence there is not a direct comparison of the modern and native types.

TABLE 6

COMPARISON OF LITTER SIZE IN MODERN AND NORTH CHINA NATIVE SOWS

Type of Sow	Number of Litters	Number of Pigs			Percent Weaned	Number Weaned per Litter
		Born	End 1st Week	Weaned		
Modern	26	262	201	179	68. 3	6. 88
Native	22	307	259	236	76. 9	10. 73

TABLE 7

COMPARISON OF FEED UTILIZATION DURING LACTATION IN MODERN AND NORTH CHINA NATIVE SOWS AND THEIR LITTERS

Type of Sow	Number of Litters	Gain per Litter	Feed per 100 Catties Gain
		(Catties)	*(Catties)*
Modern	16	233. 3	417. 2 ± 18. 67
Native	15	277. 4	322. 3 ± 10. 22

Data on performance of various breeds and crosses during the feeding period are summarized in tables 8, 9, and 10. Carcass data were obtained on some of these groups, and certain interesting items from these data are summarized in tables 9 and 10.

615999—45——10

The rate of growth in the native pigs was slower than that in either the modern purebred or the modern × Chinese groups that were fed. Native pigs required about 281 days from birth, on the average, to reach market weight (about 180 catties), while crossbred pigs (sired by modern boars and out of native sows) required about 208 days, or 73 days less to reach the same weight. Differences in efficiency of feed utilization were not nearly so marked. In the groups for which data are summarized in table 9, the two native lots required the largest amounts of feed per unit gain, while in the groups summarized in table 10, the two native groups fall within the range of the crossbred groups.

The dressing percentages available for various groups, including some animals in addition to those listed in previous tables, are summarized in table 11. The native groups fall within the range of the purebreds and crossbreds.

Some definite differences appear in the mode of fat deposition. The data for caul and kidney fats are summarized in table 12. All the differences listed in the lower half of this table are significant and indicate that the native pigs put on relatively more caul and kidney fat than the modern breeds, while the crossbreds are intermediate in this respect. The few pigs from Chinhua, Chekiang, that were studied were similar to the crossbreds, caul fat averaging 4.88±0.73 catties and kidney fat 5.10±0.70 catties, but they were slaughtered at somewhat lighter weights than the other animals and at equal weights would probably have been nearer to the other native pigs.

TABLE 8

PERFORMANCE OF MODERN PUREBRED AND MODERN × NATIVE NORTH CHINA CROSSES, 1936 DATA

(Pigs were put on feed at 13 weeks, or 3 weeks after weaning. Four pigs per lot.)

Items Compared	Breeding and Sex of Pigs									
	Hampshire Female	Berkshire Barrow	Berkshire Female	Hampshire × Native Barrow	Hampshire × Native Female	Berkshire × Native Barrow	Berkshire × Native Female	Poland China × Native Barrow	Poland China × Native Female	Yorkshire × Native Barrow
Initial weight (catties)	33.3	43.4	39.3	33.6	33.4	43.2	43.0	32.4	28.0	33.3
Days on feed	140	112	126	112	154	112	112	140	154	112
Average daily gain (catties)	1.00	1.23	1.12	1.35	0.96	1.21	1.15	1.07	1.01	1.26
Feed per 100 catties gain	390.6	360.1	381.6	341.9	401.2	382.0	374.7	399.0	394.6	338.4

TABLE 9

PERFORMANCE OF MODERN PUREBRED, NATIVE, MODERN CROSSBREDS, AND MODERN × NATIVE CROSSES, 1937 DATA

(Four pigs per litter were put on feed at 13 weeks or 3 weeks after weaning.)

Items Compared	Breeding and Sex of Pigs								
	Hampshire Barrow	Hampshire Female	Hampshire Female	Native Barrow	Native Spayed Female	Hampshire × Poland China Female	Yorkshire × Native Barrow	Yorkshire × Native Female	Chester White × Native Barrow and Female
Initial weight (catties)	42.0	45.3	44.8	20.5	16.5	32.8	39.3	39.1	56.7
Days on feed	139	133	133	220	268	123	134	130	117
Final weight	180.0	187.3	174.8	177.4	175.2	183.1	180.8	186.4	181.4
Average daily gain	0.99	0.93	0.98	0.71	0.59	1.22	1.06	1.13	1.07
Feed per 100 catties gain	455.5	462.5	459.2	484.6	540.3	403.3	415.2	352.8	465.6
Number slaughtered	4	3	4	2	4	4	4	4	4
Dressing percent [a]	72.3	71.8	70.9	72.4	74.5	74.0	71.8	69.5	76.9
Length of small intestine [b]	65.1	53.5	61.0	51.4	42.5	56.5	54.9	54.5	53.1
Caul fat, percent [a]	1.75	2.0	1.87	3.73	4.53	1.68	2.42	1.92	2.43
Kidney fat, percent [a]	1.73	2.07	1.57	3.94	5.54	1.94	2.45	2.14	2.40
Back fat (cm.) [c]	5.12	5.20	4.10	4.65	4.99	5.12	6.00	5.83	5.88

[a] Percent of live weight.
[b] In Chinese feet. Foot = ⅓ meter.
[c] At top of shoulder.

TABLE 10

PERFORMANCE OF NATIVE AND MODERN × NATIVE CROSSBREDS

(Pigs were weaned at 10 weeks and put on feed at 13 weeks. Eight pigs per lot. Barrows and gilts are included in all lots. Females were spayed in the native lots only.)

Items Compared	Breeding of Pigs							
	Native	Native	Hampshire × Native	Berkshire × Native	Berkshire × Native	Yorkshire × Native	Duroc-Jersey × Native	Duroc-Jersey × Native
Initial weight (catties)	20.9	30.8	41.3	42.4	42.2	38.2	53.1	40.4
Days on feed	192	188	115	122	126	120	94	126
Final weight	179.6	174.7	180.4	177.8	178.5	179.5	180.0	183.1
Average daily gain	.77	.77	1.21	1.11	1.07	1.18	1.35	1.13
Feed per 100 catties gain	395.2	374.9	384.6	396.1	396.8	364.6	359.9	360.3
Number slaughtered	6	([c])	8	8	7	7	8	([c])
Dressing percent [a]	71.1	--------	77.1	74.9	74.4	72.4	72.2	--------
Length of small intestine [b]	47.1	--------	55.4	49.8	49.1	54.3	66.9	--------
Caul fat, percent [a]	3.49	--------	3.26	2.75	2.48	2.12	2.67	--------
Kidney fat, percent [a]	4.55	--------	3.14	3.24	2.76	2.20	2.70	--------
Back fat (cm.)	3.75	--------	6.75	6.26	7.50	6.21	4.40	--------

[a] Percent of live weight.
[b] In Chinese feet. Foot = ⅓ meter.
[c] Data lost in evacuation of Nanking.

TABLE 11

AVERAGE DRESSING PERCENTAGES OF VARIOUS GROUPS OF HOGS

(Based on warm carcass weight)

Breeding of Pigs	Number of Pigs	Dressing Percent and Standard Error
Hampshire	12	72. 7 ± 1. 04
Berkshire	7	78. 8 ± 0. 25
Native (Chiangpei)	42	73. 3 ± 0. 43
Native (Chinhua) [a]	6	72. 1 ± 0. 58
Hampshire × native	15	76. 9 ± 0. 51
Berkshire × native	20	75. 1 ± 0. 76
Chester White × native	4	77. 1 ± 0. 19
Yorkshire × native	18	72. 5 ± 0. 60
Duroc-Jersey × native	8	72. 1 ± 0. 76
Poland China × native	7	76. 1 ± 0. 60

[a] Butchered at lighter weights than other groups.

TABLE 12

SUMMARY OF DATA ON CAUL AND KIDNEY FAT IN CARCASSES FROM VARIOUS GROUPS

Group	Average and Standard Error	
	Caul Fat	Kidney Fat
	(catties)	*(catties)*
Modern breeds	3. 06 ± 0. 20	3. 71 ± 0. 27
Modern × native	4. 66 ± 0. 09	5. 18 ± 0. 12
Native (Chiangpei)	7. 01 ± 0. 41	8. 63 ± 0. 44
Difference between:		
Modern and crossbred	1. 60 ± 0. 22	1. 47 ± 0. 29
Native and crossbred	2. 35 ± 0. 42	3. 45 ± 0. 40
Modern and native	3. 95 ± 0. 45	4. 92 ± 0. 51

Back fat was significantly thicker in native and crossbred pigs than in the modern group, but the crossbred and native pigs did not differ significantly. The data are summarized in table 13. The few Chinhua, Chekiang, pigs, which were slaughtered at somewhat lighter weights, had a thickness of back fat of 5.66±0.32 cm., or nearly as thick as that of the crossbreds and Chiangpei, Kiangsu, natives.

Thickness of skin was greatest in the native, lowest in the modern, and intermediate in the crossbred pigs. The data are summarized in table 13. In the Chinhua, Chekiang, pigs, thickness of skin was similar to the modern pigs, averaging 0.2 cm.

In an effort to determine fundamental reasons for differences in growth rate between modern and native pigs, certain organs were weighed or measured. The heart and liver were heaviest in the crossbreds and lightest in the natives. The means and differences are shown in table 13. The small intestine was longest in the modern breeds and shortest in the native. The means and differences are shown in table 14. Length of the large intestine was similar in the various groups and the small differences were not significant.

TABLE 13

Summary of Data on Thickness of Back Fat and Skin, Taken at the Top of the Shoulder

Group	Average and Standard Error	
	Back Fat	Skin
	(*cm.*)	(*cm.*)
Modern	4. 47±0. 21	0. 20±0. 003
Modern × native	5. 70±0. 14	0. 28±0. 011
Native	5. 83±0. 26	0. 33±0. 021
Difference between:		
Modern and crossbred	1. 23±0. 26	0. 08±0. 012
Native and crossbred	0. 13 (not significant) [a]	0. 05±0. 024
Modern and native	1. 36 (significant) [a]	0. 13±0. 021

[a] Actual standard error calculated but now unavailable owing to loss of records.

TABLE 14

Comparison of Development of Heart, Liver, and Small and Large Intestines in Modern, North China Native, and Crossbred Pigs

Breeding of Pig	Average Weight and Standard Error		Average Length and Standard Error	
	Heart	Liver	Small Intestine	Large Intestine
	(*catties*)	(*catties*)	(*catties*)	(*catties*)
Modern	0. 49±0. 02	2. 36±0. 13	61. 3±1. 44	15. 2±0. 46
Modern × native	. 52±0. 01	2. 62±0. 06	54. 9±1. 45	16. 1±0. 13
Native	. 46±0. 01	1. 92±0. 07	43. 1±0. 50	15. 3±0. 40
Difference between:				
Modern and crossbred	0. 03±0. 02	0. 27±0. 14	6. 4±2. 04 [a]	0. 9 [b]
Modern and native	. 03±0. 02	. 44±0. 15 [a]	18. 2±1. 52 [a]	. 1 [b]
Crossbred and native	. 06±0. 02 [a]	. 71±0. 09 [a]	11. 8±1. 54 [a]	. 8 [b]

[a] Significant differences.

[b] These differences were not significant.

Work in Chengtu. Native swine used were of two types, the rather large black pigs found in the vicinity of Neichiang, Szechwan, and the "bispectacled" white pigs typical of the Yungchang, Szechwan, area. The rations used in feeding trials were balanced according to the same standard used at Nanking, but tankage was not available and all the protein was of plant origin. Berkshire, Chester White, Yorkshire, and Poland China boars were used to sire crossbred pigs, and the data on performance of these and native pigs are summarized in table 15. All the crossbred groups grew at a more rapid rate and their gains were somewhat more efficient. No material differences appear between the natives and the crossbreds in length of the small intestine and in deposition of fat, the native litters falling between the extremes observed in crossbreds in most of these measures, the only exceptions being caul and kidney fat in which the Yungchang native pigs showed the highest levels of any of the groups.

Bristles have been an important source of income. It is not possible to predict the future importance of bristles but much depends on developments in the synthetic bristle industry. Native Chinese hogs have longer and stiffer bristles than the modern hogs of the United States, and the methods of management result in a much longer period for bristle growth before the hogs are slaughtered. Under pre-war

TABLE 15

PERFORMANCE OF NEICHIANG AND YUNGCHANG NATIVE PIGS, AND CROSSES WITH MODERN BREEDS

(Five pigs per litter were used.)

Items Compared	Breeding of Pigs							
	Yungchang Native	Neichiang Native	Berkshire × Yungchang	Berkshire × Neichiang	Yorkshire × Yungchang	Yorkshire × Neichiang	Chester White × Yungchang	Poland China × Neichiang
Initial weight [a]	33.4	30.3	29.3	37.2	30.6	36.0	28.6	31.1
Days on feed	224	230	181	186	161	202	179	171
Final weight	183.6	187.1	181.5	183.2	183.1	180.8	182.2	182.3
Average daily gain	0.67	0.59	0.82	0.78	0.95	0.72	0.86	0.88
Feed per 100 catties gain	487.0	524.5	448.8	420.2	438.4	474.2	449.6	425.2
Length of bristles (cm.) [b]	3.24	3.04	3.38	2.82	4.61	2.85	3.03	3.40
Number slaughtered	5	3	4	4	5	5	5	5
Dressing percent	65.9	61.8	64.3	67.6	73.2	64.8	64.8	65.1
Length small intestine [c]	62.0	64.5	63.9	60.4	59.4	54.5	65.5	66.7
Caul fat, percent	2.46	1.96	1.01	2.05	2.08	2.14	1.76	1.47
Kidney fat, percent	2.94	2.42	2.08	2.39	2.74	2.10	1.75	1.97
Back fat (cm.)	5.12	5.93	6.35	4.05	4.88	5.42	7.28	6.50

[a] In catties. Catty = ½ kilo, or about 1.1 pound.
[b] Bristles over withers.
[c] In Chinese feet. Foot = ⅓ meter.

TABLE 16

GROWTH OF PUREBRED, NATIVE, AND CROSSBRED PIGS UNDER NATIVE FARM CONDITIONS

Breeding of Pigs	Number		Average Initial		Average Final	
	At Beginning	At End	Age	Weight	Age	Weight
			(days)	*(catties)*	*(days)*	*(catties)*
Native	4	1	100	30.0	380	186.0
Native	3	2	120	41.0	288	86.0
Berkshire	4	2	100	27.0	352	70.5
Berkshire	2	2	160	40.5	384	100.5
Hampshire	4	3	100	25.8	380	119.3
Large Yorkshire	7	4	154	41.5	348	128.6
Chester White×native	4	2	120	31.0	288	129.0
Duroc-Jersey×native	4	3	100	34.5	380	201.7

conditions white bristles were more valuable than black, and the Yungchang hogs were one of the chief sources of white bristles. Data on bristle length at the end of the feeding period were obtained on the animals for which performance data are presented in table 15. Data on bristle length are summarized in the same table. They are inconclusive, owing to small numbers. The crossbreds compared favorably and in one case exceeded the native animals in length. However, bristles on the crossbreds were not as stiff as those on the native animals.

In the various feeding trials summarized in preceding portions of this paper, the animals were maintained under experiment-station conditions, with management and feeding practices as comparable with those used in the United States as the circumstances permitted. Conditions under which Chinese pigs are commonly raised and fed are quite different, and these conditions have already been described.

A limited test was made of the performance of some native, modern purebred, and crossbred pigs under experiment-station conditions until they weighed from approximately 100 to 150 pounds. They were then distributed to farmers to be fed in their own way. Weights were taken at intervals of four weeks until the pigs were approximately 1 year old. The data are summarized in table 16. Obviously, the numbers are too limited to permit definite conclusions. Study of the data does not indicate definite superiority of any of the three general groups (native, modern purebred, and crossbred) used in this study. Much further work is required to determine the adaptability to Chinese farm conditions of the modern breeds and crosses between these and native types.

Chickens

Types

The chickens of China are a heterogeneous mixture for the most part. Large and small types, feathered and clean-legged types, crested and non-crested birds, and a variety of plumage colors and comb types can be found in almost any area. Little organized breeding has been done. Chickens are frequently carried from one village to another for presentation to friends and relatives as gifts. Many of these remain alive and interbreed with the birds of the village to which they are taken. Consequently there has been much mingling of local types, and most places have quite a mixture.

FIGURE 73.—*Top:* Farm flock of chickens and two geese near Yungchang, Szechwan. *Bottom:* Cocks of Kundeling, Tibet.

FIGURE 74.—*Left:* Chickens of Nanking (photographed by T. Y. Hsu, Chinese Ministry of Agriculture). *Right:* A trio of native chickens in Shansi. The average farm in any area has only a few birds.

Observations and discussions with workers from various portions of the country do indicate, however, that certain localities have types that are more or less distinct. A type that is similar to the Cochin is raised in the Yangtze Valley. It is also called Shantung chicken, owing to a belief that it originally came from that province. The birds are yellow in color and have feathered legs. They may or may not have a crest. Mature birds are reported to weigh up to nine catties. A type is raised in Kiangsu and possibly in other areas which is called Langshan. It is said to have originated in the northeastern provinces. It is black, has feathered legs, and weighs up to eight catties at maturity. A type called Yuchi or oil chicken is found in the vicinity of Peiping. There is considerable variation in size (two or three, up to six catties) and some are crested and some non-crested. They are quite fat, hence the name oil chicken. They are also reasonably good egg producers, apparently, for it is reported that by some selection within the native non-crested type, egg production has been raised through selection by James A. Hunter to an average of about 150 per year according to Tong (1943), the best birds producing as high as 200 eggs. The native non-crested hens are reported to average about 110 eggs, while the crested average 60 to 70 eggs per year.

Silkies are said to be found to some extent in Kiangsi and may exist in other provinces. They weigh one to one and one-half catties. Feathers are white and the skin is black. Bones are also black. A type called Sheokwang is found in Shantung that is reputed to produce large eggs. Tong (1943) reports that eight hens from purchased eggs averaged about 120 eggs in a laying year, and that the average egg weight was approximately 80 grams. Common colors are black, spotted gray, and yellow. Hens weigh about five and cocks six catties. Another type is called Hsiaoshan. This type is found in Chekiang and is uniformly yellowish in color. They tend toward a dual-purpose type, being reasonably good for both egg and meat production. Mature hens are reported to weigh about five catties.

There are no doubt other types that may be sufficiently distinct to justify listing as separate from the heterogeneous mixtures usually found, but there is not sufficient information at hand to justify further listings at this time. The chickens of the northwest differ somewhat from those in the farming regions of central and southern China. In the latter regions, the chickens tend toward the large meaty type, while in the northwest the tendency is toward the smaller egg-laying type, with some that tend definitely toward the game type.

Little information is available concerning the egg-producing ability of the chickens of various types and in various areas. In addition to the figures already cited, Tong (1943) quotes the results of three experiments conducted by James A. Hunter, Raymond T. Moyer, and himself which indicate that the native chickens studied averaged 50 to 70 eggs per year. The Japan-Manchukuo Yearbooks (1938–39–40), reviewed by Wang (1942), state that the chickens of Manchuria are mostly of native type, that egg production is less than 100 eggs per bird per year, and that large-type native birds average about 60 eggs, but no actual data are given. Trials conducted by Lingnan University indicate that the native chickens around Canton produce about 60 eggs per year (Tong, 1943). These records were obtained with birds that were kept in enclosures and were fed a standard improved ration.

Ko (1940) studied the size of eggs on the Chengtu market and also the size of eggs produced by local native chickens and certain purebreds maintained at Central University. The average egg sizes (in grams) observed were as follows: Chengtu market eggs, 49.79; native eggs, 51.57; Langshan eggs, 57.46; Leghorn eggs, 52.63; Rhode Island Red eggs, 53.82; and Barred Plymouth Rock eggs, 51.31.

Some data have been collected by Hsu and Fan (1943) on egg production, egg size, age at first egg, and growth in certain native types. These data are summarized below: numbers are limited so naturally cannot be considered adequate to represent the average characteristics of the various types, but they do indicate something of the level of production. The three groups studied were black birds that were native to Yishan in Kwangsi Province, Silkies, and birds that were native to Kiangsi Province. Data on egg production for a period of one year (June through May, inclusive) are summarized below:

Group	*Number Birds*	*Average Number Eggs*
Black Yishan	14	90
Silkie	7	55.7
Kiangsi native	6	90

Egg size was determined for some birds of these groups. The averages are indicated below:

Group	Egg Weight (grams)
Black Yishan	40.9
Silkie	42.7
Kiangsi native	35.5

The average for the black Yishan natives was obtained from 14 hens and the range in average for each hen was from 36.3 to 44.3 grams. Age of first egg was determined on nine black Yishan and nine Kiangsi native hens. The range for black Yishan natives was 132 to 225 days, with an average of 165.4, while the Kiangsi natives ranged from 132 to 177 and averaged 152.6 days. Weight at 200 days for 13 black Yishan birds was 1.14 kilos and for 12 Kiangsi native birds was 0.82 kilo.

Management and Feeding

The aggregate number of chickens in China is large, but they do not occupy a large place in farming operations on the average farm. They function more as scavengers and utilizers of table and other refuse. A survey made by Buck (1937) indicated that the most common number of chickens on the individual farm is three, and numbers commonly observed range from 1 to 20.

Housing provided for chickens is not elaborate, and many types are found. Bamboo coops or large baskets with a small opening at the top may be used for housing at night or where it is not convenient to allow the chickens to run loose during the day. Wooden crates with slat bottoms may be used in areas where lumber is plentiful. A type of adobe house is used in Hopeh. This consists of a cell about 3×3×3 feet. It has only one opening, about 1.5 feet from the ground on one side and large enough for the birds to go in and out. Poles are laid in the adobe as the house is built, about one foot from the ground, which serve as roosts. The small opening is closed with a board or other cover at night. There is no provision for ventilation or for cleaning out the manure that collects under the roosts. An improved modification of this type has been developed which has openings for ventilation and for cleaning. Some of the openings are covered with wooden lattice set into the adobe at the time of building, so that there is ventilation at night even though the birds are closed in for protection. In some sections and on some farms no housing is provided, the chickens simply being allowed to run about and find a place to roost wherever they may choose for the night.

The chief feeds for chickens are whatever they pick up from foraging about the farm, and scraps from the table. Some grain is fed, the kind depending upon the grain raised in that area. There are reports that maggots are raised for feeding poultry in some sections of Fukien and Hunan, but this is apparently practiced to only a limited extent.

There is no well-organized marketing procedure. Eggs and live or dressed birds are taken to the local marketing centers and sold by the growers on market days, or sold to retailers who resell them to the consuming public. Some purchasing is done in the country by middlemen who in turn take the eggs or birds to market.

FIGURE 75.—Chickens and other poultry and eggs are sold in the cities and villages on market days. *Top:* Poultry market on top the city wall in Yungchang, Szechwan, showing the bamboo baskets in which fowl are commonly carried to market in South China. *Bottom:* Egg market in Yungchang, Szechwan.

Dried eggs were an important export product prior to the present war. Most of these were prepared at points along railroads in the northern portion of China and in Shanghai.

Breeding

Perhaps the best characterization of breeding practices with chickens in China is to say that nature has simply taken its course. With the exception of a few institutions that have tried in a small way to improve the native types by selection, there has been no attempt to improve the native stock. In fact, selection may have been against such a character as large egg size, owing to a general tendency to sell large eggs for human food and reserve small eggs for hatching.

Cocks usually run with the hens throughout the year. Some castrating of males is practiced to produce capons for marketing.

The method of incubating eggs in hatcheries is different from that developed in other countries. The natives of Tungchangfu, Shantung, have practiced a method for generations, passing it on from father to son, that will serve to illustrate the principle used. Hatcheries employ these operators on a contract basis, the company advancing enough money to maintain the operator's family during his absence and paying his and his apprentice's expenses for the hatching period. The operator guarantees the hatchery a hatch of 85 percent of the fertile eggs. At the end of the hatching season, the operator receives $1/20$ of the net profit. The eggs for hatching are purchased by trained buyers from farms in the vicinity. The incubator is a *kang* or earthern bed, under which a fire is built for heating. Each kang is about 6 feet wide, 10 feet long, and 3 feet high. Two decks are built above this, one about 4 feet from the kang and the second about 4 feet from the first. The kang is covered with a sheet of felt and the eggs laid on top of it. The eggs are then covered with a cotton quilt. The temperature is about 110° F., but no thermometers are used by the native operators, who determine the temperature by touching the egg to some sensitive part of the skin such as the eyelid. The eggs remain on the kang for one week, after which they are candled by sunlight and the fertile ones placed on the first deck above the kang. Here they are covered with a quilt and kept for one week, at a temperature of about 105° F. They are then candled a second time and all the live ones placed on the top deck and covered with a quilt. The temperature is about 95° F. The entire room is also at about this same temperature. When the first peep is heard the quilt is removed from the top deck. As the chickens hatch they are removed and kept in the incubating room to dry. If business is rushing, the chicks may be sold to peddlers as soon as dry. Otherwise, they are kept for about 24 hours and then sold.

Incubation usually begins about the middle of July.

Another method of incubation which is practiced in some areas is to provide the heat with hot rice. This is more commonly used with duck eggs but is also used in some areas for the incubation of chicken eggs. There are local variations in the details of procedure but all are similar to that outlined below.

A wooden barrel is used which is just a little larger at the top than at the bottom. This barrel is approximately 87 cm. high and 60 and 57 cms. in diameter at the top and bottom, respectively. The barrel is lined with about 3 to 3.5 cm. of straw and a stronger layer of paper is placed inside this. Approximately 13 cm. of rice hulls are placed in the bottom of the barrel and covered with paper. A layer of heated rice is placed on top of the paper. Eggs are placed in fish-net sacks, about 60 to the sack. Two sacks of eggs are then placed in the barrel, one around the outside and one in the center. A layer of warm rice about 6.6 cm. deep is placed on the eggs. This process is repeated until the barrel is filled to the proper level, and a barrel, of the size indicated above, holds 14 layers, or a total of 1,680 eggs. A layer of paper is placed on top the last layer of eggs and about 3.3 cm. of warm rice poured on top of this. If the weather is cold a second layer of paper and rice may be added, and if quite cold a third layer of paper and a cotton pad will be placed on top of this. The rice is heated to 45–50° C. and placed in a special container. A total of about 120 catties is needed for each batch of eggs. Two barrels are used, and the eggs are shifted from one barrel to the other with each change of rice, so that 60 catties must be heated each time. When put in contact with the eggs it is about 40° but operators determine if it is at the proper temperature by placing it against the eyelid.

Approximately 2 catties of pork fat or tea-seed oil is mixed with the 120 catties of rice to make it smooth. This is added gradually during the first two or three days of incubation. The first change of rice is made after two hours, then changes are made every eight hours. The changes will be more frequent than this in cold weather and less frequent in warm weather. With each change the eggs that were in the top go to the bottom of the barrel, and in each layer the sack that was on the outside goes to the inside. Turning of the eggs is accomplished by the process of picking up the fish-net sacks and putting them down again in the second barrel. The temperature of the eggs is usually about 35° at the time a change of rice is made. The process of changing requires about 10 minutes. Chicken eggs are kept in the hot rice for 14 days. After this they are producing enough heat themselves so that a layer of hot rice is required only at the bottom and top of the barrel. Later on the top layer may be omitted. The eggs are moved every eight hours or so, as usual. At the end of 16 or 17 days the eggs are removed from the barrel and put on a traylike arrangement. Each tray may be about 2 meters × 1.25 meters and be about 13.5 cm. deep. It is made of wood. The sides of the tray are lined with several layers of paper. A layer of rice hulls about 3.3 cm. deep is placed on the bottom and covered with a straw mat. The eggs are placed on top of this mat and covered with several layers of paper. The number of layers depends on the weather, and if it is cold, a quilt may be added. The eggs are checked at frequent intervals to make sure that the proper temperature is being maintained. They remain on this tray until they hatch. Eggs are candled at the 5th or 6th day and again at 10 to 14 days. Final candling is usually done at 18 or 19 days.

Various observers report that White Leghorns, Rhode Island Reds, Barred Plymouth Rocks, Black Orpingtons, White Wyandottes, and other breeds have been introduced into China. Many of these have been brought in by missionaries, and various institutions have made importations. No definite information is at hand on the times and extent of these importations, and likewise there are very few data at hand to show the results obtained in grading-up operations. Considerable data are no doubt available and should be brought together and assembled in sufficiently complete form to indicate what improvement has resulted from the use of the various breeds, both in egg and meat production. Observations by various workers indicate that grading up with such breeds as the White Leghorn, Rhode Island Red, and Barred Plymouth Rock resulted in increased egg production as compared with the native types. Results with the White Leghorn may not have been so satisfactory from the standpoint of meat production, and this result would be anticipated owing to the differences in the purebred types.

In the Japan-Manchukuo Yearbooks (1938–39–40) it is stated according to Wang (1942) that Leghorns, Rhode Island Reds, and a Japanese breed had been introduced into Manchuria, but no data are given to indicate the results obtained.

Data have been accumulated on a limited number of hybrids between Single-Comb White Leghorns and two native groups in preliminary work by Hsu and Fan (1943). The results are summarized below. The ranges in age at first egg and the average age in each group are shown below:

Group	Number of Hens	Range	Average
		(days)	*(days)*
White Leghorn	4	174–252	214.0
Leghorn × Kweilin (F_1)	7	174–277	211.9
Leghorn × Kweilin (F_2)	9	137–222	165.3
Leghorn × Hunan (F_1)	6	163–236	205.3
Leghorn × Hunan (F_2)	6	151–190	167.0

The Kweilin group were native of that area in Kwangsi Province, and the Hunan group were native to the area around Hengyang in that province. The average egg sizes observed in three groups were as follows: White Leghorn, 53.6 grams; Leghorn × Kweilin (F_1), 37.9; and Leghorn × Hunan, 43.7 grams. Average weights of birds in these groups at 200 days were as follows:

Group	Number	Average Weight
		(kilos)
White Leghorns	12	1. 106
Leghorn × Kweilin (F_1)	12	1. 037
Leghorn × Kweilin (F_2)	20	1. 075
Leghorn × Hunan (F_1)	8	1. 341
Leghorn × Hunan (F_2)	11	1. 123

Tong (1943) gives the following results of some grading-up work done at Minghsien College:

Breeding	Age at First Egg	Average Egg Weight	Average Number Eggs in Laying Year
	(days)	*(liangs)**	
6 Leghorns	185	1. 7	182
6 Hybrids	227	1. 35	118
6 natives	302	1. 15	62

*Liang is approximately 1⅛ ounces.

Ducks

Types

There are several types of ducks in China that are recognized as being distinct. These are described briefly below.

Peiping ducks are white in color, with yellow beaks and feet. The females weigh about five catties and the drakes six to seven catties.

Nanking ducks are either white with yellow beaks and feet like the Peiping ducks, or they are brownish gray with a speckled pattern. The majority are of the latter color. They weigh about one catty less than the Peiping ducks and have less development of the breast.

Muscovy ducks are produced in Sikong and in parts of Szechwan, Kwangtung, and Kwangsi Provinces, and in some other places. They are rather large, weighing somewhat more than the Peiping ducks. Colors are either white or black, with red pigmented areas around the base of the beak and the eyes. Black is the predominant color.

The ducks of the rice-growing areas which do not fall in the above classes might also be listed as a type. They are small, weighing two to three catties when mature. The most common color is brownish gray, in a speckled pattern. Some other colors are also seen.

Ko (1940) studied the size of duck eggs in three types and found the averages to be as follows: Chengtu, 57.75 grams; Nanking, 63.65 grams; and Peiping, 72.82 grams.

Management and Feeding

Ducks, like chickens, are used to a considerable extent as scavengers, but the business is somewhat specialized. The most common number on farms, according to Buck (1937), is 2 and commonly observed numbers varied from 1 to 25, but in some instances there are as many as 100 or even 500. The business of preparing them for market is more specialized also.

In Szechwan the common practice is to feed ducks first at about 2 days of age, when they are given cooked rice and some rice bran. As soon as they are able to scavenger for themselves, they hunt their own food in ponds and rice fields with occasional supplemental grain feeding, until they are about 2 months old. Bands of them are then collected by professional herders who fatten them as they are driven to a central marketing point. The usual practice is for three men to handle

615999—45——11

FIGURE 76.—Ducks and geese are used largely as scavengers. *Top:* A flock of ducks scavenger along a stream bed near Yungchang, Szechwan. *Bottom:* A pair of geese on a farm near Yungchang, Szechwan.

a herd. Two guide the ducks, using long bamboo poles with tassels at the ends. The third man carries the portable bamboo house and cooking utensils and a portable fence for the ducks. They move slowly along, after the rice harvest, allowing the ducks to feed on grain that has fallen in the fields. In return for this free "pasturage" the farmer gets the manure they leave behind. Herds range from several hundred up to a thousand or more. This practice of herding to market is also common in other areas, and in the Nanking area the droves are often larger, numbering as many as 3,000 ducks. The ducks in these droves are, for the most part, raised by farmers who keep only a relatively small number during the growing period and sell them to the drivers who make a business of fattening them on the way to market.

Methods vary somewhat in different areas. In Kwangtung Province, for example, it is reported that ducks are fed small fish and fish refuse in addition to rice products. Some small frogs are also caught from the rice fields for feeding. Ducks are housed in temporary sheds along streams in this area. Farmers collect their eggs along the stream bank.

Some ducks are also kept in boats along the river. They forage along the stream banks and pick up refuse in the stream and are kept on the boat for safety at night.

Some of the fattening of ducks is done by restaurants, who prepare them for their special trade. The unfattened ducks are purchased, placed in small pens, and force fed on table refuse and on grains and grain by-products until fat enough for use on the restaurant tables.

Ducks may be moved to market in droves, as has been described earlier, or they may be taken in small numbers, either alive or dressed. Some are sold alive to restaurants, where they are fattened before being used. Many ducks are consumed soon after killing, but a considerable number are preserved.

Methods of preservation vary in different areas. A few of the common practices are noted below.

The viscera are removed through the rear end of the bird. They may then be dried and salted, after which they are marketed in grocery stores. Gizzards are dried separately and sold as a delicacy.

A special method of drying is used in the Canton area. The duck is spread out on a bamboo frame, salted, and dried in caves. It is reported that this is done during one season when the wind is from the south. Hearts are also salted and sold separately in some areas. Livers may be salted and dried for use in making a special kind of sausage. Whole ducks are also dried and smoked.

Duck eggs may be sold in the fresh state, or they may be treated either with salt or so treated as to produce "lime" eggs. Whites of fresh eggs are sometimes used for the treatment of fish nets, so they will last longer when used in salt water. Yolks of such eggs are salted, dried, and sold for table use.

Ducks are usually picked when wet. The feathers are dried and used for making pillows.

Breeding

Little effort has been made to improve the productivity of ducks, either for egg or meat production. Various types predominate in various areas, as outlined above, but these appear to have been chance development rather than the result of purposeful improvement programs.

The method of hatching duck eggs that is more commonly used is essentially the same as the method described for chickens, in which warm rice was used as a source of heat. In addition to barrels, boxes are used as containers. These may be built with four compartments, each of which would hold about 1,000 eggs. As many as 100 eggs may be placed in one net. The procedure is essentially the same as that with chicken eggs, except that duck eggs are kept in the rice until the 21st day, then removed and allowed to incubate on trays by their own heat. The last candling of duck eggs is made on the 26th or 27th day.

Peddlers of ducklings sometimes buy unhatched eggs and let them hatch out while they are en route to market.

On warm, sunny days the sun may be used as a source of heat. Eggs are simply laid out in the sun for 20 to 30 minutes or until they are warm enough. The operator keeps rotating them to maintain the proper temperature. After two weeks of such sunning during warm weather, they may be laid on a quilt, covered with another quilt, and the "eggs incubate the eggs". It is claimed that this method gives 90 and sometimes as high as 95 percent hatchability of the fertile eggs.

The supply of eggs for hatcheries may be obtained from small farmers, either direct from the farms or at market places. Duck farms also supply eggs to hatcheries, usually on a contract basis, whereby they receive a certain amount of money and a set number of ducklings for each 100 or other definite number of eggs provided. Duck eggs are also hatched by hens on farms but not by ducks.

It is claimed that some well-trained operators can determine the sex of ducklings either by the feel of the copulatory organ or by looking at it. This distinction is said to be more easily made in newly hatched birds than later on. Similar methods have been developed in other countries for chickens; the skill of these men should be tested scientifically, and if their method is accurate, the facts should be so recorded that all who can use the technique profitably can learn it.

Some workers claim that a cross between Muscovy ducks and the brownish-gray ducks of the rice areas produces a very desirable bird that grows rapidly and has good-quality meat.

Geese

Types

All the geese are of essentially the same type, with the exception that there are two colors, white and gray. The white geese are claimed to be a little larger than the gray ones, on the average. The white type predominates in northern China and in Szechwan, while most of the grays are found in southern China.

Little information is available on egg production. Some Szechwan farmers claim that their geese produce 75 to 100 eggs per year. Such information as is available indicates that geese in Chekiang and Kiangsu produce 40 to 50 eggs per year on the average.

Studies of the rate of growth in geese by Tong (1943) at Minghsien College have shown that goslings grow very fast from birth to the end of the sixth week, after which they grow slowly until a rather constant weight is reached. At 4 months the ganders weighed about 8 catties and the geese about 6.5 catties.

Management and Feeding

Geese are kept by farmers and village and city dwellers for the eggs and meat produced, and also as house watchers. When any strange persons come near the house the geese make noise to warn the occupants, thus functioning as guards against thieves. There is also a belief in some areas that snakes will keep away from houses where geese are kept, because the snakes are afraid of geese manure. This belief is based on the unfounded idea that if the manure gets on the body of the snake the flesh will rot.

The management and feeding of geese is similar to that for ducks, except that there is less specialization with geese. The usual number of geese on farms, according to Buck (1936), is two and the number is seldom very large. The geese usually forage for themselves in streams, ponds, and around the farm. Sometimes algae are collected from ponds for feeding. In addition to such feed as they pick up, they may be fed table scraps, left-over cooked rice, rice bran, and vegetable tops.

Geese may sleep wherever they choose around the house or may be housed with the chickens in a bamboo or other type of coop. However, they are usually left free, so that they will be more apt to detect prowlers.

Geese are usually sold alive at marketing centers on market days. Those sold are usually ganders and geese that have outlived their usefulness for egg production. Roast goose is sold in restaurants, and some birds are dried and salted. The dried ones are usually sold in grocery stores.

Eggs are usually sold while fresh at the local marketing centers for about 25 percent more than the prevailing price for chicken eggs. Eggs are sometimes salted and sometimes cured in brewery swills before selling.

Breeding

The breeding situation is very similar to that for chickens and ducks. Little effort has been made to improve either the meat-producing or egg-producing qualities, and nature has taken its course in the mating process.

Most goose eggs are hatched by hens on the farms and the hatching is seldom done by the goose. Some eggs are hatched by duck-egg hatcheries in the manner already described.

Improving Chinese Livestock

Breeding

Most of the farm livestock and poultry in China are of rather primitive, unimproved types. Some of the existing types are meeting the needs of the country reasonably well but others need to be improved in certain of their functions so that they can make greater contributions to national welfare. Other types are in need of improvement in all their productive characteristics. There are three main methods by which improvement in productive capacity could be brought about. These are: (1) Selection within the native types, (2) grading up with already improved types or breeds from other countries, and (3) development of new types out of animals that are graded only part of the way up to the purebred type.

Selection within native types has certain advantages. There is then no problem of selection and importation of foreign stock, and the native types are already adapted to the local environment, which may not be true of types brought in from other countries. The disadvantage of this method is that progress will be much slower than with grading up with already improved types, provided the improved types meet China's needs and are adapted to the environment.

Available information from breeding experiments indicates that the variation in most economically important characters is owing to a large extent to non-hereditary factors, and that, in an improvement program, not more than 10 to 30 percent of the progress reached for by selection of parents can be realized in their offspring. For example, if a character like grease weight of fleece is found to be 20 percent heritable, and a group of ewes and rams are selected for breeding that average one pound above the population from which they were selected, their offspring will on the average produce only about one-fifth pound more wool than the original population. Thus, in a population of sheep that produce 1.5 to 2.0 pounds of grease wool, on the average, progress by selection to a 6- or 8-pound level would require many generations of careful selection and breeding.

Grading up has the obvious advantage of making possible rapid improvement, through the use of improved males on large numbers of native females. It is the logical choice of a method in all cases where there are existing types in other countries that meet China's needs and which are apt to be able to perform satisfactorily under Chinese conditions. If this method proves satisfactory, foreign breeds or other improved herds need be maintained only as sources of males to continue the grading-up program, while the bulk of the animals in the country would become high grades as the program progressed.

In many cases it may be found that the foreign purebred animals are not able to perform satisfactorily under existing environmental conditions. Grades that are carrying a certain amount of improved breeding may withstand the environment satisfactorily and produce at higher levels than the native stock. In such cases, the obvious procedure is to determine the extent to which improved breeding can be introduced, then develop a new type from animals with this breeding background.

The breeds that can be used most successfully in China, either for grading up or for use in developing new types from grades, will remain unsettled until much more preliminary work is done in many portions of the country to determine the adaptability and response of various types and breeds and their crosses with the native stock. All existing information concerning characteristics and adaptability of foreign breeds and similarities of conditions under which they are produced to those existing in sections of China where they would be used should be evaluated when selecting types and breeds to try out in China. For example, the Red Sindhi and Sahiwal cattle of India should be useful for grading up of cattle in the rice area of China, owing to their adaptability to a warm environment and their relatively high milk production as compared with the native cow. Similarly, the Murrah and related breeds of water buffalo of India, that produce a considerable amount of milk, should be useful for developing dual-purpose water buffalo in China by the grading-up process.

The Brown Swiss breed of cattle, which was developed in mountainous and rather cold areas, should be adapted to use for grading up the draft cattle of the Yellow River basin to improve their milking ability. This breed might also be used for crossing with yak to produce larger and more productive *pien niu.* Many such examples could be given for other classes of livestock. However, these should be sufficient to indicate the approach that should be used in selecting breeds to be tested in China. Final selections of breeds that should be used in grading up or in developing new types on a large scale would have to await results of breeding and performance tests under the conditions in which they would be used.

FIGURE 77.—Members of the First National Animal Husbandry Conference, held under the auspices of the Chinese Ministry of Agriculture, September 15 to 30, 1943, in Yungchang, Szechwan, to discuss existing and post-war animal husbandry problems. *Front row, left to right:* Tsun San Chiang, Kao Hui (Louis) Lu, Tung Shen, Wu Chi Tsai, Ralph W. Phillips, Shao Chun Cheng, Fan Chun Chang, Chung Hsien Wu, Tze Chun Liang. *Back row, left to right:* Chi Shien Chang, Wen Tien Ma, Wen Ho Peng, Chen Chiu Huan, Qui Sheng Tong, Yao Tsong Chen, I Shu Chen, Shun Yin Chang, Chong Hsien Sun. *Not in picture:* Tseng Ying Hsu, Yei Zen Tang.

Inbreeding is a useful tool in livestock improvement, once improved types are reasonably well established, but it seems probable that it should be avoided so far as is possible in the early stages of an improvement program in China. Inbreeding in the improved stocks that are used for grading up, etc., would not be objectionable, particularly if it had been used as a means of increasing homozygosity in outstanding strains. But its use in developing improved types by grading up and then selecting to fix a new intermediate type should be avoided until the new type has been fixed to some extent by selection. To do this a large number of sires should be used in initial stages of the programs.

Hybridization will continue to be useful as a means of producing mules, and also of *pien niu* providing satisfactory methods can be developed for utilizing the *pien niu* females effectively in continued breeding operations.

Environment

There are two basic truths that the animal-husbandry specialist would emphasize to his workers. The first is that agriculture involves not only the tilling of the soil and raising of crops but also animal production. Plant and animal production are both parts of agriculture and should be so considered. The second is that grassland is a valuable asset and in many places should be left as such instead of being placed under cultivation. In many large regions of China grass is the basic resource, and the only satisfactory and economic method of harvesting a crop off such land is to let livestock harvest it.

One frequently hears of plans to greatly expand the human and livestock populations of the grassland areas, so that these so-called "waste lands" can be more fully utilized. Such programs cannot hope to succeed except in a few very limited localities. In many areas the numbers of livestock should be reduced, and in only a few areas could any increases be made at present. With the development of adequate methods of range management and control of grazing, the total livestock population could gradually be increased, and with this some increase of human population could be made. But such developments should be preceded by adequate development of the range and by development of the small amounts of available adjacent lands that are suitable for cultivation, so that they would supplement the range-livestock industry rather than conflict with it as is now the case.

The view that livestock numbers are at or near the maximum for existing range conditions is supported by all trained observers who have visited the grasslands and with whom the writer has discussed this problem. It is also the conclusion of Ross (1936), who studied the problem in Manchuria and in parts of Inner Mongolia. In western Manchuria and Inner Mongolia severe droughts are not uncommon, and these when they occur may accentuate the effects of any tendency toward overstocking. Cressey's summary (1934) of the potentialities of the Central Asiatic steppes should be cited in connection with this problem: "In this waste all life struggles incessantly against the restrictions imposed by the climate and only those plants and animals survive that can adjust themselves to the meager resources available. The possibilities of occupation, shelter, food, fuel, and clothing are all laid down by nature. The one resource is an uncertain supply of short dry grass."

In this statement, Cressey shows the limitations of the range lands of Central Asia but at the same time it should be recognized that the range lands of the world constitute a major resource. Not only do they furnish sustenance for the vast numbers of livestock grazing thereon but they also constitute the watersheds for hydroelectric and irrigation developments downstream. On this grassland domain, under present economic and market conditions, where the wealth of the inhabitant is measured in livestock numbers rather than long-time net production, there is little economic incentive for livestock producers to expend either financial or human effort to improve production efficiency. When markets for wool, meat, hides, and animals become good enough, when currency becomes stable, and when marketing agencies are recognized as being service organizations for the benefit of producer and consumer, then progress in the grasslands can be expected. The economic pressure will then force the present residents to action; otherwise outsiders will move in and take their place.

Just as soon as the livestock producer's objective becomes net surpluses of animal products, he will be ready to think of animal efficiency rather than just the ability of an animal to withstand the rigors of an unaided and unmanaged "mother nature". Since increasing efficiency of individual production is linked with decreasing resistance to malnutrition, adequate amounts and quality of feed will be recognized as a necessary factor for advancement in animal production. Nutrition will be recognized as being as important as, if not more important than, the vaccinating needle. On range or grasslands, the primary problem for advancement is one of a year-round supply of feed. The lush growing season, often short, produces the low-cost feed. The dry, cold winter season is one of high cost—or one of livestock losses. China today cannot pay for high-cost winter feed so takes the loss just as the "early-day" livestock producer of the range lands of the western United States did. With dependable, appreciative markets and ways for reaching them, the grassland livestock man will be in a better position to see the need for obtaining a better forage cover on the range. He will then be ready to make use of modern grassland management. He will see the necessity for a winter feed supply, whether it be standing grass on deferred ranges or hay on nearby cultivated or meadow lands. Civil authorities and government planners will then see and recognize the interdependence of range lands with wintering grounds. Range lands are only good so long as they have a complementary winter feed source, that absolute prerequisite to efficient livestock production.

Today, there is a basic competition between the livestock producers and the farmers that encroach on the edge of the grazing lands. These farmers move in and develop small subsistence farms in the valleys where the nomads have been accustomed to grazing their animals during the winter and early spring months. Thus the nomad is pushed out of grazing land that is needed for the most critical seasons of the year. Much of this land is marginal, and the result is that the farmer makes a poor living and the nomad has a poorer living than he had before. As this process goes on the nomads suffer more, and the country as a whole loses products that it badly needs. The logical development in such areas is to devise a system of agriculture in which the available crop land is used to grow supplemental feed for the range livestock so that they can be carried safely through severe winter storms and the early spring when feed is very scarce on the range itself, plus some food for the livestock owners themselves. This would eliminate much of the winter death loss and the females would be better able to maintain pregnancy and nurse their young satisfactorily in the spring and summer. The result of such a development would be greatly increased supplies of wool, meat, and leather, and other products for use in China, and an improved market for manufactured products from other parts of the country. The possibilities of better utilization of grassland livestock products will be discussed later under marketing, but all this is involved in improving the environment for livestock production in the grasslands. Along with developments of this type should go careful study of range conditions and methods of managing those ranges so that maximum production can be obtained.

The basic competition between the nomad and the farmer was recognized in Manchuria and Inner Mongolia by Ross (1936) who stated: "So far as the present volume of wool production in Manchukuo and Inner Mongolia is concerned, by far the greater part of this comes from the purely pastoral zones and very little from the agricultural areas. In the border territories between these two zones, there is likely to be increasing competition between the demands of an agricultural and pastoral economy, so the tendency will be for sheep grazing to be increasingly restricted rather than the reverse." It also appears that this same competition between farmer and nomad in Inner Mongolia was partially responsible for the break between Outer Mongolia and China at the time of the fall of the Ching dynasty, owing to resentment among the Mongols over repeal of laws which had until a short time previously prevented colonization of grazing land by Chinese farmers. The fall of this last dynasty gave them opportunity to break away from a government that had not recognized their basic needs and rights.

Experience proves that a man's wealth cannot be measured by how many animals he owns. The important point is not how many animals he owns but how much marketable product he is able to harvest from those animals each year. In many cases it is possible to increase production by reducing numbers to the optimum carrying capacity of the range.

Immediate improvement in the producing capacity of the grassland areas can be brought about by improving the hereditary capacity of the animals to produce wool and other products, as has already been emphasized. Increases in numbers should await development of feed supply adequate to maintain in livestock a nutritional level sufficient to meet the requirements of animals bred for higher levels of production than those now maintained.

In the crop-producing areas of China, the major effort is now expended on the production of grain for human food, and with the existing density of population it is necessary that this emphasis continue. But there is room for increased emphasis on livestock and for improving the environment of livestock without interfering with the production of human food. In grassland areas, crop production should supplement livestock production, while in farming areas livestock should supplement crop production. Careful studies should be made of the possibilities of producing more forage crops, especially legumes, and root crops that could be used for livestock feed so that the diets of working animals on these farms can be improved. Similar studies should be made of the possibilities of improving the method of feeding swine and poultry. Some grain might be diverted from human consumption for these purposes and to support milk production, either by animals that are used for work or by goats that are kept for this purpose, in order to increase the amounts of animal protein and calcium in the diets of farm people. There must be provision for sufficient good-quality forage and other feeds if these animals are to provide a sufficient supply of milk for the farm family. Marketing of some crops through livestock is essential to maintenance of soil fertility by return of manure to the land and is especially important in China if the existing practice of using night soil is to be eliminated in the interest of public health.

The discussion up to this point has dealt with improving the diet of animals. There is room for improvement of other phases of the environment also.

The Chinese farmer does not normally give much thought to his livestock. The nomads have a closer tie with livestock but this group also has not developed sufficiently improved methods of handling their animals. There is urgent need for dissemination of knowledge of improved methods of breaking, training, and handling horses so that working and riding animals can be used more easily and efficiently. This will be especially important as improved strains are developed and made available for use. The overloading of many draft and pack animals is noticeable. Overloading in farm work is not so common, but in commercial transportation by carts and on pack animals an overload is the rule.

There is need for careful study of the equipment used by draft, pack, and saddle animals (horses, donkeys, mules, camels, cattle, yak, *pien niu*). Horse collars, for example, are particularly in need of attention since they are so short they choke the animal, and the surface that touches the shoulder is usually of cloth which is rough and cannot be kept soft and clean, so collar sores are frequent. Methods of multiple hitching should also be devised that make more efficient use of horsepower than those that are now in common use. Improved types of saddles are also sadly needed. Carts and other equipment should be studied carefully to determine how more useful and efficient vehicles can be made. The Peking cart, for example, is a very useful vehicle and is of a size that one animal can conveniently handle. But the basic idea has been expanded into the large commercial cart that is loaded with 1½ to 1¾ tons of material. Regardless of how many animals are hitched to such a load, the brunt of the burden falls on the wheel horse or mule, since the entire weight of the load is thrown on this animal with every move and lurch of the cart.

The common practice of forcing animals to work while they are reined up tightly cannot be too highly condemned. The reason for this practice is not clear, but it may be owing to fear, from the lack of good horsemanship, which prompts drivers to rein up their animals tightly so that they can be more easily controlled.

All such practices and the lack of efficient equipment are contributing factors to the working animals' environment, and if an optimum environment is not provided the hereditary possibilities of the native animals, and especially of improved stock when it is available, cannot be realized.

More efficient utilization of draft animals should be possible through the development of large fields and the communal working of farms. Studies of Buck (1937) have shown that water buffalos are used, on the average, 42 days per year on farms. In addition to greater utilization of these animals for work, there is the possibility of dual use for work and milk.

Equipment for other types of livestock is also important. In sheep, for example, it is important that producers have good shearing equipment. Existing types of shears are very primitive. If they could be exchanged for efficient ones, the quality of the wool marketed would be improved, owing to getting the full length of the fibers and avoiding so many second cuts.

Housing of livestock is also in need of attention. This is especially true in grassland and other cold areas where there is need for shelter for protection from winter storms. Such shelters need not be elaborate or expensive. Agricultural engineers should study this problem and devise types that can be constructed easily from materials native to the various areas. This problem should also be studied in farm areas so that more adequate types can be developed for all classes of animals. Well-being of the animal, including sanitation, as well as sanitation for the farmer, also should be considered when developing such plans.

There is an extremely large task awaiting agricultural engineers who have an understanding of livestock problems, in the development of improved equipment and housing, along the lines which have been indicated briefly above.

The problems of diseases and parasites must also be considered in the improvement of the environment. Chinese workers are now conscious of the disease problem but are apt to overemphasize the importance of the serum bottle and the injection needle. Production of virus, serum, etc., is vital to the maintenance of livestock health, and there is no intention to minimize its importance here. However, there is need for relatively greater emphasis on sanitation, on management practices that help to control parasites, and on improved nutrition so that animals will have sufficient reserve energy to better withstand disease and parasite attacks. The vaccinating needle can never correct malnutrition.

Marketing

The producer of livestock must be assured a fair price and an appreciative market for his livestock and livestock products if he is to be expected to make real efforts to improve the quality of those animals and products. So long as there is no distinction made in the returns from products of different qualities, and so long as the producer is cheated by profiteering middlemen and manufacturers, there is little hope of having the farmer and the nomad really interested in livestock improvement.

A fair price at the market is also essential if livestock and their products are to reach the consuming public at reasonable prices and become a real contributing factor in national well-being and economy.

In order to guarantee such a market, some well-organized type of control over marketing procedures by the central government is essential. An honest control is just as essential as an honest marketing system.

The development of monopolies for the handling of animals and their products should be avoided, so far as this is possible. A government monopoly may be necessary in certain situations, but such a monopoly must be of a benevolent nature, considering only the interests of the producer and the consumer, and be operated without "squeeze" if it is to serve the best interests of the nation. Monopolies that are operated for profit by the government (either national or provincial), or that are operated jointly by the government and bankers or by the government and manufacturers, are not apt to fall in the category of benevolent monopolies, and such organizations could best be avoided in the future.

A market should exist primarily as a service organization, not as a primary source of income from an industry, if the industry is to be developed to its fullest and most useful extent.

Standardization of products is essential to orderly marketing. These standards should be of a nature that is easily understood by both producer and consumer, so that transactions can be carried out on the basis of these standards and all parties to the transaction have a clear understanding of the quality or merit of the product involved.

Rigid control of dishonest marketing practices, such as the adulteration of milk with water and the pumping of water into hog and beef carcasses and into poultry, is essential. Such practices are now reported to be quite common. The enactment and enforcement of pure-food laws would eliminate these and some similar practices.

Development of livestock marketing will be closely associated with the development of transportation. This is especially true of the marketing of animals and products from the grasslands. The whole problem of how best to develop the marketing from these areas should be carefully studied by experts in livestock marketing. The results of these studies should be considered in the development of highways, railroads, and river transportation, so that the best interests of the livestock producers and the nation as a whole will be served.

The importance of close association between the economy of the grassland and the economy of the more densely populated areas of the country has already been stressed. The resources of the grassland should be harnessed to provide meat, milk, wool, hides, and other products that are needed in the diet and for manufacturing purposes in other parts of the country. Adequate means must be developed to make these products available. In turn, the more densely populated areas will find a market for certain food materials and manufactured products in the grassland areas. Increased exchange of products between these two areas should be encouraged, in the interest of a better-balanced national economy.

Education on the efficient utilization of livestock products is important, not only to help develop markets but to help the consumers make the most efficient utilization of and derive the greatest benefit from these products. Education on the value of milk in the diet is an example of this, which would also serve as a stimulus to more rapid improvement of milk-producing animals. Research on methods of cutting meat and poultry and on methods of cooking these so that greatest good can be obtained from them should be an important phase of livestock development in the future.

Conclusion

China has a large and varied livestock population, but no comprehensive description of the various types and the conditions under which they are produced and used has been available. This publication has been prepared in an attempt to fill that need. It is intended as a reference work for agricultural research workers, teachers, students, and others interested in livestock improvement in China.

Owing to wartime conditions and other limitations, the information is incomplete. In addition to the authors' own observations, material has been drawn from many sources, but much more information must be assembled before a complete description of China's livestock can be had. This work should, therefore, not be considered a completed study but rather a beginning or a base upon which other studies can be built, so that complete information on China's livestock types and their productivity will be available.

A complete discussion of ways of improving livestoek is outside the scope of this report. A more detailed discussion of this problem in China is given by Phillips (1944). Careful study of the many problems should be made by breeding, nutrition, and livestock-marketing experts and a program developed on the basis of these findings.

It is hoped that this volume will serve four main purposes:

(1) Be useful as a reference volume on Chinese livestock until a more complete work can be prepared.

(2) Indicate to Chinese agricultural workers and political leaders the magnitude of their livestock industry and its place in national economy.

(3) Develop greater interest in livestock improvement among agricultural and political leaders in order that the livestock industry can make an even greater contribution to national welfare.

(4) Stimulate greater interest in animal industry among students, in order that more of them may see the great opportunities for service in this field and undertake livestock improvement as their work.

615999—45——12

Literature Cited

ALLEN, GLOVER M.

1940. The Mammals of China and Mongolia. Natural History of Central Asia XI: 621–1291 (Genus *Sus Linnaeus* 1114–1129).

BUCK, JOHN LOSSING.

1937. Land Utilization in China. University of Chicago press.

BURNS, R. H., JOHNSTON, A., and CHEN, W. C.

1940. The Improvement of Chinese and Other Carpet Wools. Jour. of Textile Institute 31: T37–T48.

CHANG CHI HSIEN (Ministry of Agriculture and Forestry).

1943. Unpublished data. (Ref. translated from Chinese.)

CHANG SUNG YIN (National Szechwan University).

1943. Unpublished data. (Ref. translated from Chinese.)

CHANG TAI HUA.

1941. The Products of the Camel. Jour. An. Husb. and Vet. Sci. 2: 30. (Ref. translated from Chinese.)

CHEN I. SHU (Ministry of Agriculture and Forestry).

1943. Unpublished data. (Ref. translated from Chinese.)

CRESSEY, GEORGE BABCOCK.

1934. China's Geographic Foundations, Ed. 1. McGraw-Hill Book Co., Inc., New York.

CROSS, H. E.

1917. The Camel and Its Diseases. Bailliere, Tindall and Cox, London.

DICKINSON, FRANK (West China Union University).

1943. Unpublished data.

GOROSENKO, JA. L.

1936. Mongoljskaja koza (The Mongolian Goat). Domasnie Zivotnye Mongolia, 175–194. Abstracted in Animal Breeding Abstracts 5: 156 (1937).

HUAN CHEN CHIU (Ministry of Agriculture and Forestry).

1943. Unpublished data. (Ref. translated from Chinese.)

HUANG YI SHENG (Kansu Incorporation for Agricultural Improvement).

1943. Unpublished data. (Ref. translated from Chinese.)

HSU MU CHUN (China Wool Textile Factory).

1943. Unpublished data. (Ref. translated from Chinese.)

HSU TSENG YING.

1940. Dressing Percentages of Szechwan Fat Hogs. Published by Dept. An. Husb. and Vet. Sci., National Central University. (Ref. translated from Chinese.)

———

1940. Report on Tzetung (Szechwan) Cattle Show. Published by Szechwan Provincial Bureau for Agricultural Improvement. (Ref. translated from Chinese.)

———

1940. One Year of Swine Research in Szechwan. Published jointly by National Central University and Szechwan Provincial Bureau for Agricultural Improvement. (Ref. translated from Chinese.)

HSU TSENG YING.
1941. Summary Report on Swine Research. Jour. An. Husb. and Vet. Sci. 2: 35–39; 73–80. (Ref. translated from Chinese.)
——— (Ministry of Agriculture and Forestry.)
1943. Unpublished data.
——— and CHANG LUNG CHI.
1941. The Use of Individual Feeding Method for Feed and Breed Comparisons. Jour. An. Husb. and Vet. Sci. 1: 296–311. (Ref. translated from Chinese.)
——— and CHIANG YU FANG.
1940. Szechwan Bristles. Published by Dept. An. Husb. and Vet. Sci., National Central University. (Ref. translated from Chinese.)
——— and FAN PAO CHEN (Ministry of Agriculture and Forestry).
1943. Unpublished data. (Ref. translated from Chinese.)
——— and PENG WEN HO.
1939. Second Year's Report on Swine Research (1937-38). Published by Dept. An. Husb. and Vet. Sci., National Central University. (Ref. translated from Chinese.)
——— ———
1941. Progeny Testing of Szechwan Sows. Jour. An. Husb. and Vet. Sci. 4: 1–15. (Ref. translated from Chinese.)
——— ———
1942. The Effect of Diet on Growth and Reproduction in Neichiang Gilts. Chinese Jour. of An. Ind. 1: 24–34. (Ref. translated from Chinese.)
——— ——— ——— (Ministry of Agriculture and Forestry).
1943. Unpublished data. (Ref. translated from Chinese.)
——— ——— ——— and FENG KWANG CHAO.
1938. Report on One Year's Swine Research (1937). Published by Dept. An. Husb. and Vet. Sci., National Central University. (Ref. translated from Chinese.)
KISLOVSKY, D.
1938. The Domestic Animals of Mongolia (Review of Russian book edited by J. Lus). Jour. Heredity 29: 27–32.
KO FU CHIANG.
1940. Average Egg Weight of Poultry Eggs in Chengtu in Comparison With Those From Purebreds. Jour. An. Husb. and Vet. Sci. 1: 29–32. (Ref. translated from Chinese.)
LATOURETTE, KENNETH SCOTT.
1934. The Chinese, Their History and Culture. Ed. 2, rev. The Macmillan Co., New York.
LEVINE, C. O.
1920. Swine, Sheep and Goats in the Orient. Jour. Heredity 9: 117–124.
LIANG TZE CHUN (Chungking Dairy).
1943. Unpublished data. (Ref. translated from Chinese.)
LU KAO HUI (West China Union University).
1943. Unpublished data. (Ref. translated from Chinese.)
——— and LU KAO CHAO (West China Union University).
1943. Unpublished data. (Ref. translated from Chinese.)
MCMEEKAN, C. P.
1938. Shape of the Growth Curve as a Controlling Factor in Conformation and Anatomical Composition of the Animal Body. Amer. Soc. Animal Prod., Proc. 337–341.

MOYER, RAYMOND T., and Q. S. TONG.
1944. Unpublished data from Minghsien College, Taiku, Shansi.
MUNZ, H.
1937. Observations on Some Wool Samples From Northeastern Asia. Commonwealth of Australia, Council for Scientific and Industrial Research, Pamphlet No. 69.
PHILLIPS, RALPH W.
1944. Livestock Improvement in China. Published in English and Chinese by the Chinese Ministry of Agriculture and Forestry, Chungking.
——— and KNAPP, BRADFORD, JR., HEEMSTRA, LOUIS C., and EATON, ORSON N.
1943. Seasonal Variation in the Semen of Bulls. Am. Jour. Vet. Res. 4: 115–119.
——— and ZELLER, J. H.
1943. Sexual Development in Small and Large Types of Swine. Anat. Res. 85: 387–400.
ROSS, I. CLUNIES.
1936. A Survey of the Sheep and Wool Industry of North-Eastern Asia. With Special Reference to Manchukuo, Korea and Japan. Commonwealth of Australia, Council of Scientific and Industrial Research, Pamphlet No. 65.
SAITO, MICHIO.
1936. Some Macroscopical, Microscopical, and Chemical Investigations on Mongolian Sheep Wool. Report of Inst. Scientific Research, Manchukuo, 1 and 29–62.
SHA FENG PAO (Ministry of Agriculture and Forestry).
1943. Unpublished data. (Ref. translated from Chinese.)
SZECHWAN BUREAU FOR AGRICULTURAL IMPROVEMENT.
1939. Two Years' Breeding Records on Swine. Published by Szechwan Bureau for Agric. Improv., Chengtu. (Ref. translated from Chinese.)
TANG YEI ZEN (National Central University).
1943. Unpublished data. (Ref. translated from Chinese.)
TONG QUI SH'ENG (Minghsien College).
1943. Unpublished data. (Ref. translated from Chinese.)
WANG HSIEN KAI.
1941. Anatomy of the Camel. Jour. An. Husb. and Vet. Sci. 2: 6–12; 44–52. (Ref. translated from Chinese.)
WANG PEI CHIEN.
1942. Condition of Animal Husbandry in Recent Years in Manchuria and Japan (a translation of material in Japan-Manchukuo Yearbooks for 1938, 1939, and 1940. Chinese Jour. Animal Industry 1: 205–220. (Ref. translated from Chinese.)
ZAWADOWSKY, M. M.
1931. Zebu-yak Hybrids. Jour. Heredity 23: 297–313.

Titles in the Modern Chinese Economy Series

THE LATE IMPERIAL PERIOD

(late nineteenth and early twentieth centuries)

1. Ball, Samuel. **An Account of the Cultivation and Manufacture of Tea in China.** London, 1848.

2. Chen, Jerome. **State Economic Policies of the Ch'ing Government, 1840–1895.** New York, Garland, 1979.

3. Edkins, Joseph. **The Revenue and Taxation of the Chinese Empire.** Shanghai, 1903.

4. Fortune, Robert. **Three Years' Wanderings in the Northern Provinces of China.** London, 1847.

5. King, Paul. **In the Chinese Customs Service: A Personal Record of Forty-Seven Years.** London, 1924.

6. Parker, E.H. **China: The History, Diplomacy, and Commerce from the Earliest Times to the Present Day.** London, 1917.

7. Parsons, William Barclay. **An American Engineer in China.** New York, 1900.

8. Wilkinson, Endymion Porter. **Studies in Chinese Price History.** [Ph.D. Dissertation, Princeton University, 1970.] New York, Garland, 1979.

9. Wright, Stanley. **Kiangsi Native Trade and Its Taxation.** Shanghai, 1920.

THE AGRARIAN ECONOMY

10. Buck, John Lossing. **Three Essays on Chinese Farm Economy.** New York, Garland, 1979.

11. **Economic Facts.** [published by the Department of Agricultural Economics of the University of Nanking, Chengtu.] September 1936–January 1946. Nos. 1–53. Four volumes.

12. Lamb, Jefferson D.H. **The Development of the Agrarian Movement and Agrarian Legislation in China.** Shanghai, 1934.

13. Roll, Charles Robert, Jr. **The Distribution of Rural Incomes in China: A Comparison of the 1930s and 1950s.** [Ph.D. Dissertation, Harvard University, 1974.] New York, Garland, 1979.

14. Tang, Chi-Yu. **An Economic Study of Chinese Agriculture.** Cornell, 1924. [Ph.D. Thesis.]

15. Wagner, Wilhelm. **Die Chinesische Landwirtschaft.** Berlin, 1926.

FOREIGN TRADE AND INVESTMENT

the Rise of the Modern State, and Their Impact upon the Hinterland

16. Arnold, Julean. **Commercial Handbook of China.** Washington, 1919, 1920. Two volumes.

17. Burrill, Harry R. and Raymond F. Crist. **Report on Trade Conditions in China.** Washington, 1906.

18. Chen Han-seng. **Industrial Capital and Chinese Peasants: A Study of the Livelihood of Chinese Tobacco Cultivators.** Shanghai, 1939.
Bound With
The China Institute of Economic and Statistical Research. **A Study of the Rural Economy of Wuhing, Chekiang.** Shanghai, 1939.
Bound With
Chin Chien Yen. **Wool Industry and Trade in China.** Tientsin, 1937.

19. **The Commercial, Industrial and Economic Situation in China.** London, 1926, 1927, 1928, 1930.
Bound With
Moser, Charles K. **Where China Buys and Sells.** Washington, 1935.

20. Wagel, Srinvas R. **Finance in China.** Shanghai, 1914.

TWENTIETH-CENTURY ECONOMIC DEVELOPMENT

21. The American Presbyterian Mission. **The Isle of Palms: Sketches of Hainan.** Shanghai, 1919.

22. China International Famine Relief Commission. **Herr Raiffeisen Among Chinese Farmers.** Peiping, 1930.

23. Duncan, Robert Moore. **Peiping Municipality and the Diplomatic Quarter.** Peiping, 1933.

24. Hosie, Alexander. **Manchuria: Its People, Resources and Recent History.** London, 1904.

25. Institute of Pacific Relations. **Economic Trends and Problems in the Early Republican Period.** 1931.

26. Kann, Edward. **The History of China's Internal Loan Issues.** Shanghai, 1934.
Bound With
B. Lyn Yu Shen: **China's Currency Reform: A Historical Survey.** Shanghai, 1941.

27. King-yi Hsu. **Political Mobilization and Economic Extraction: Chinese Communist Agrarian Policies During the Kiangsi Period.** New York, Garland, 1979.

28. Kraus, Richard A. **Cotton and Cotton Goods in China, 1918–1936.** [Ph.D. Dissertation, Harvard University, 1968.] New York, Garland, 1979.

29. Myers, Ramon H., ed. **Selected Essays in Chinese Economic Development.** New York, Garland, 1979.

30. Remer, C.F. **Readings in Economics for China.** Shanghai, 1933.

31. Sakatani, Baron Y. **Manchuria: A Survey of Its Economic Development.** [Revised by Grover Clark, prepared for the Division of Economics and History of the Carnegie Endowment for International Peace.] 1932.

32. Woodhead, H.G.W. **Extraterritoriality in China: The Case Against Abolition.** Tientsin, 1929.
Bound With
Fong, H.D. **The Growth and Decline of Rural Industrial Enterprise in North China.** Tientsin, 1936.
Bound With
Ho, Franklin L. and Hsien Ding Fong. **Extent and Effects of Industrialization in China.** Tientsin, 1929.

LABOR AND THE ECONOMY

33. Epstein, Israel. **Notes on Labor Problems in Nationalist China.** New York, 1949.

34. Fang Fu-an. **Chinese Labour: An Economic and Statistical Survey of the Labour Conditions and Labour Movements in China.** Shanghai, 1931.

35. Wagner, Augusta. **Labor Legislation in China.** Peking, 1938.

THE WARTIME ECONOMY AND POSTWAR PROBLEMS

36. Ch'ao-ting Chi. **War-time Economic Development in China.** New York, 1939.

37. Dawson, Owen L. "Food." No. 6 in **Studies of Relief and Rehabilitation in China.** August 1943.
Bound With
Dawson, Owen L. "Rehabilitation Problems in Agriculture." No. 14 in **Studies of Relief and Rehabilitation in China.** August 1943.
Bound With
Problems of Economic Reconstruction in China Series. H.D. Fong, K.Y. Lin, and Tso-Fan Koh, **Principles of Postwar Economic Reconstruction in China** (Part I) and **Capitol Stock in China** (Part II). Quebec, China Council, 1942.

38. Phillips, Ralph W., Ray G. Johnson and Raymond T. Moyer. **The Livestock of China.** Washington, 1945.

39. Taylor, George E. **Japanese Sponsored Regime in North China.** New York, 1939.

40. United Nations Relief and Rehabilitation Administration. **UNRRA in China.** Washington, 1948.
Bound With
Ray, J. Franklin, Jr. **UNRRA in China: A Case Study of the Interplay of Interest in a Program of International Aid to an Underdeveloped Country.** New York, 1947.
Bound With
Committee on Foreign Affairs, House of Representatives. **Text of Proposed China Aid Bill and Background Information on Economic Assistance Program for China.** Washington, 1948.